D0904806

Instructional Systems Development for Vocational and Technical Training

instructional systems development for vocational and technical training

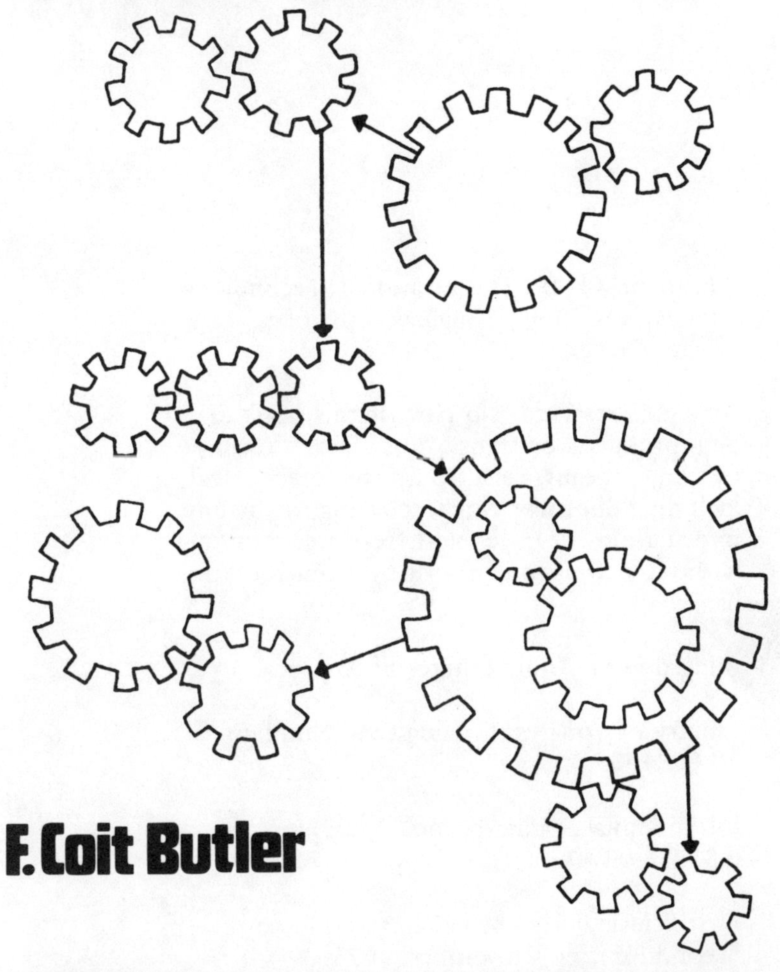

F. Coit Butler

EDUCATIONAL TECHNOLOGY PUBLICATIONS
ENGLEWOOD CLIFFS, NEW JERSEY, 07632

Printed in the United States of America.

Library of Congress Catalog Card Number: 70-168490.

International Standard Book Number: 0-87778-027-7.

First Printing: March, 1972.
Second Printing: November, 1973.
Third Printing: September, 1976.
Fourth Printing: December, 1979.

ABOUT THE AUTHOR

F. Coit Butler is director of curriculum research and development at the New England Resource Center for Occupational Education. He has worked more than 20 years in pioneering research and development in instructional systems technology, with the United States Air Force, the American Institutes for Research, and other organizations. As a senior research scientist at the American Institutes for Research, he directed efforts of two curriculum development teams involved in cross-media research and development of instructional systems which incorporated the most efficient and effective combination of new methodology, technology and content. These two applied research projects involved the development, validation, installation, and evaluation of instructional systems for both vocational training and academic studies.

ACKNOWLEDGMENTS

This book results from my good fortune of having known and worked with a host of creative people over the last twenty years. I have merely tried to distill that experience in a manner that will be useful to others not so fortunate. Among the many who, directly and indirectly, contributed so much to this book, I am most indebted to the following: John V. Patterson, Joseph A. Tucker, Robert M. Gagné, Leslie J. Briggs, Norman L. Crowder, Arthur A. Lumsdaine, Arthur R. Steiger, Gabriel D. Ofiesh, Stephen A. Church, Roland C. Ewing, James W. Altman, Edward J. Morrison, Vivian (Hudak) Guilfoy, J. William Ullery, Robert E. Pruitt, Glen E. Neifing, James B. Thrasher, Glen P. Nimnicht, Thomas F. Gilbert, Joseph I. Lipson, Lee D. Brown, and Lawrence P. Creedon.

Finally, I must acknowledge the patience and forbearance of my wife, Alva, whose help made it all possible.

AUTHOR'S PREFACE

The Systems Concept

The "systems approach" has been very successful in dealing with the tremendously complex design and development problems involved in the production of defense weaponry and space vehicles. Systems engineering is more than just the application of multiple-disciplines, a scientific team effort, although that is a vital part. The process involves the accurate identification of the requirements and problems, *the setting of specific performance objectives,* the application of logic and analysis techniques to the problems, the development of methods for the solution of the problems, and *the rigorous measurement of this product against the specific performance objectives.*

The same systems engineering concepts so successfully applied to the development of hardware can also be applied to vocational training course development problems. As a result, *an empirical methodology for the analysis, design, development, and evaluation of vocational curriculum* has been evolving. Training course design decisions now can be based, to a considerable degree, on analysis of actual student performance data. Each step in the development process can be empirically tested and validated against these performance data. Thus, the training course developer has a tool for determining the validity of training objectives, content, sequence, method, media, and achievement. In addition, training managers and school administrators now have an empirical basis, derived from performance data, for assessing the efficiency and effectiveness of training, rather than "subjective" opinion alone.

The Purpose of the Book

The instructional systems development concept, process, and product are described in detail. The first five chapters contain a general discussion of the instructional systems development concept, the learning theory behind it, the elements of an instructional system, and the systems development process. The discussion is intended to give the user the needed background for understanding and appreciating the principles underlying the procedures. The remaining chapters and the appendices provide detailed, how-to-do-it guides for the systems development process.

The procedures described in this book furnish training managers, school administrators, and course developers with an empirical basis for judging the validity of training course objectives, content, sequence, method, media, and achievement. Although the main purpose of this book is to furnish vocational curriculum developers with concise, step-by-step instructions and decision guides to the procedures involved in instructional systems development, it can also serve as a quality control document for training managers, school administrators, and instructors.

F. Coit Butler

TABLE OF CONTENTS

Appendices **Page**

Instructional Systems Development for Vocational and Technical Training

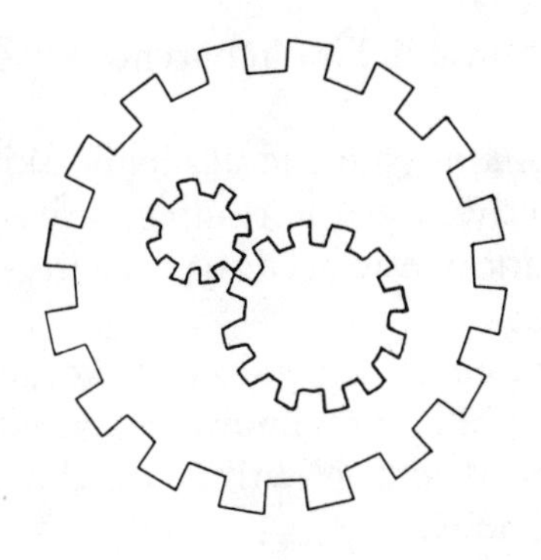

CHAPTER I

PERSPECTIVES ON OCCUPATIONAL EDUCATION

Occupational education involves a complex, interrelated set of goals, activities, organization, facilities, and functions that must operate as an integrated and coordinated system to turn out a finished product—the graduate who is fully prepared for the challenges of the urbanized, industrialized society of today and tomorrow. A vocational training program has to be thought of as a system with inputs and outputs that are controlled by *very specific behavioral objectives and performance criteria.*

There are four broad behavioral objectives and performance criteria for occupational education, all closely interrelated and each supportive of the others. By graduation, or even at a point before graduating, whether from a secondary or post-secondary school, the student should be capable of demonstrating the following behavioral characteristics:

- *The minimum,* specific vocational skill and knowledge needed at the entry level to a distinct job cluster.

- *The minimum,* specific physical, emotional, and social skills and knowledge in group and individual living needed to sustain him in his entry-level job.

- *The minimum,* specific academic skills and knowledge that directly meet the reading, writing, speaking, listening, and arithmetic needs of his entry-level job.

- *The maximum*, generalizable vocational, social, and academic skills and knowledge needed for his future advancement and growth in his chosen occupation and as an individual.

Outwardly, the emphasis may appear to be on job training; and in a way it is, but only in the broadest sense. Actually, the *primary concern must always be the student's physical, emotional, and social development—his living and working habits and attitudes.* Training in specific job skills is, of course, very important in itself, but it is also probably the single most effective instrument for bringing about the desired overall personal development. *Only within the context of training for specific marketable jobs will the student develop the kind of self-discipline, self-respect, and self-direction needed for responsible, productive citizenship.*

Everything done in an occupational education program must be done in the name of, and be directly related to, specific job training. The staffing, the curriculum, the training methods, and the training facilities must all point to this end. Ideally, the training should take place within an environment that combines the characteristics of a real work situation and a learning laboratory. However, the methods and content of the formal instruction must be distinctly job-oriented rather than lecture-oriented. *The really important skills, knowledge, habits, and attitudes can best be learned in a work-like environment, not in a standard classroom setting.*

Finally, and most importantly, *the student himself must be able to see clearly the relationship between what he does in training and the job he is looking toward.* Unless there is real and obvious application for what he is asked to learn, he will rapidly lose interest. Only in a work-like, job-oriented situation is there the kind of satisfaction and sense of progress needed to motivate the student to finish what he set out to do when he first enrolled in the program.

Perhaps the most crucial problem in developing a vocational training program is to select or define objectives. Vocational education has no meaning in the abstract. Students can learn facts, principles, skills, concepts, and attitudes, but they cannot learn all of them. Education implies organized learning of some of them, and any particular selection and sequencing of learning experiences has meaning and usefulness only for certain activities. To be meaningful and useful, vocational training courses must be designed to provide each student with the opportunity to develop skills, knowledge, and attitudes that are relevant to post-graduation competence and achievement, and that are congruent with his needs, aspirations, and potential. Course content must be selected by starting with an enumeration and analysis of what competent people actually must know and be able to do in the real world. From this study of competent performances, one can derive the skills and knowledge which the student must acquire and, thus, establish an effective sequence for these learning experiences. The point is that *each learning experience is in the program because it must be there if the student is to be competent, and the justification for its presence can be demonstrated on the basis of relevance to an occupational goal.* Different goals of different students will have different educational paths. This is what is meant by relevant vocational education.

One of the tragic figures in life is the man who once was useful, but whose finely developed skills and knowledge have become obsolete. Such a man finds himself skilled, but unnecessary. Vocational education must find a way to produce graduates who can accommodate to changes in the demands of the market place. One attack on this problem is to provide experience in the skills and knowledges which are common to a family of task-related occupations and, thus, to minimize the amount of new training one must undergo when the requirements of a job change or when opportunities open up in a related area. *Analyzing the requirements of many jobs within selected broad vocational areas for common and related tasks, discloses "core" or generalizable task elements which become the basis of the occupational versatility of the graduates.*

Qualifying for an entry-level job is a necessity, but such can no longer be viewed as a final career commitment. An entry-level job should be looked upon as only the first rung up an occupational ladder leading to a career in an upward mobile series of task-related jobs. Vocational courses must be developed that will provide a broader base of skills and knowledge based on clusters of task-related jobs. Such courses must begin with basic entry-level tasks and proceed stage-by-stage through progressively higher-level job-related tasks, thus providing students with a range of exit points to match their needs and capabilities. Additionally, these job-related courses should offer an integrated curriculum that starts in high school and extends upward through two-year and four-year post-secondary technical and engineering programs. The same functional curriculum framework applies as well to non-technical, non-engineering social service career fields. In fact, there is no occupational curriculum that cannot be restructured to encompass a series of job-related courses that lead upward from relatively low-level tasks, which can be learned at the secondary level, to fully professional duties, which require years of advanced study. *Every job-related course should be open-ended in that it always leads directly to a higher-level course and its related occupation.*

No one should claim the ability to foretell the technological future sufficiently well to forecast all the skills and knowledge which will become useful or even critical in even one occupational area during the working lifetime of present students. Nor would it be wise, or possible, to educate present students in all future requirements even if we knew them. Consequently, an attempt must be made to develop in students the ability and the motivation to pursue new learning throughout life. In particular, the entire course must be built around developing the generalizable skills needed for "problem-solving" in, for example, the broad field of electronics. *Course content must be concerned with the revelation and the learning of processes as well as the acquisition of pure information and skills.*

One of the earliest empirical findings in psychology was that dependable differences exist between people. Since those early days, it has been demonstrated repeatedly that *individuals differ with respect to a great variety of abilities,* to say nothing of motivations, attitudes, and personality characteristics. Each of us takes some account of individual differences in our daily lives as we deal with other people. Yet, it appears that most people do not appreciate fully the extent of individual differences and their importance for the educational process.

It is difficult to believe that adequate provision for individual differences can be made within the framework of the traditional classroom. The teacher who is aware of students' needs, and conscientious about meeting them, will not have the time required to devote attention to each student in addition to regular class activities. As for student motivation, it is likely to be adversely affected in many instances. A relatively small segment of students, who can just "keep up" with the pace of instruction, may be stimulated by participation in class activities. The more able group is likely to be insufficiently challenged, whereas the less able group suffers from a continued sense of failure. In point of fact, *the major problem in present training programs is the failure to deal adequately with students differing in ability and progress.*

To solve this problem, instruction must be designed to fit the individual student. By individualizing instruction, learning becomes a process guided by the teacher, while remaining the responsibility of the individual student. The student's achievement becomes the standard of his learning progress, and at the same time a primary source of his motivation. *Course content must be organized into self-contained, self-instructional, self-evaluating lesson units, each of which is based on a specific task-like activity.* Each unit is headed by a list of objectives which tell the student what he is expected to be able to do after completing the assignment. The students work individually on the assignments using a carrel, work table, or whatever location and equipment are required. Self-study, individual tutoring, small-group discussion among students at a similar stage of progress, demonstrations and experiments can all be a part of individualized instruction. The key point, however, is that *students do the learning largely on their own* and student-teacher interactions do not degenerate into lecture classes. Finally, when a student completes an assignment and is satisfied that he has mastered it, he informs the instructor, who administers the appropriate performance test, scores the test immediately, and, if the student has succeeded, enters the information on the student's record card and makes another assignment.

Individualizing instruction, in the manner described or in some variation of it, is essential if *all* students are to be equipped with the skills and knowledge they need to become competent, productive, and flexible workers. Only in this way can an accommodation be made for the enormous differences in abilities, interests, and motivations of students. A truly individualized program must provide an opportunity for selection of individual goals, and for alternate routes to those goals, at a pace appropriate to the individual. Further, by arranging the sequence of learning experiences to progress stage-by-stage through a series of task-related job levels, *there is a chance to ensure that each student will leave school with some marketable skill, whatever his ability level, his aspirations, or, within reasonable limits, his time in school.*

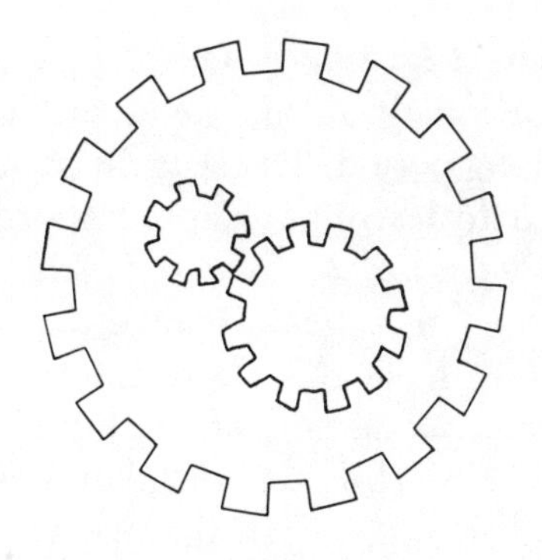

CHAPTER II

A SUMMARY OF THE FACTORS AFFECTING LEARNING

Introduction

There are, first of all, no hard and fast laws or rules for learning, but there are some general conditions that usually facilitate most kinds of learning. Also, there are exceptions to every stated principle, to every condition. There are seven general factors that, when correctly taken into account, usually encourage learning. On the other hand, if any of these seven factors is neglected or incorrectly applied, learning is usually discouraged. The seven factors are: *Motivation, Organization, Participation, Confirmation, Repetition, Application,* and *Individual Differences.*

Motivation

The learner must want or need to learn if there is to be much learning. There has to be incentive for learning to take place. There is very little accidental learning of any degree of complexity. Within reasonable limits, the greater the motivation, the greater will be the learning. Several sub-principles fall within this area.

Giving or stating to the student the goals to be achieved during the learning session will strengthen motivation. There must be *distinct learning goals.* To be motivating, the goals must be clear to the learner. Learning is favored by an early, definitive statement of the overall goal, and by specific statements of the goals for each learning period. The student should know each time what he is expected to learn; but, more importantly, he should be *told exactly what he is to do with the new knowledge or skill.* Obviously the student must be persuaded that the goal is *desirable* and must believe that the goal is worth the effort. Accordingly, he must see the goal as *attainable.* The learner must think there is a real chance for success. Thus, it is usually best to set a series of fairly easy, short-range goals when the material or task is long and complex and offers only long-term payoff.

Motivation is enhanced and sustained if the learner is told of his progress from time to time. He must know if he is succeeding or not; he must have *knowledge of progress.* If the student is having trouble, positive suggestion that he can succeed aids motivation. *Encouragement* aids motivation, whereas negative suggestion hampers motivation. Praise for correct answers or good performance is usually more motivating than reproof for poor performance.

Motivation is favored when the material is presented realistically and demonstrates the practicality of the content. *Realism* and *practicality* are most effective when the instruction clearly shows that what is being learned will actually be used on the job.

Concurrently, the learner's motivation will be strong if he expects to use the knowledge or skill in the immediate future. *Anticipated early use* strengthens the desire to learn. If the student knows that he is going to perform the same steps immediately after they have been demonstrated, he will pay more attention to the demonstration. Or, if he knows he is to be tested on the content immediately following the showing of a film, he will watch the film more closely. Frequent *exams* and *quizzes,* if correctly used, can help to sustain motivation. The learning situation must be a *challenge* if it is to be motivating.

Learning conditions have to maintain a fine balance between success and failure—if the task is too easy, the student is bored; if the task is too hard, he is frustrated. However, it is usually more motivating if he is pressed to move a little faster and a little farther

than he expects or feels he can go. If the material or task is too easy and the learner is always successful, he may lose interest because of the lack of challenge. There has to be some failure against which to measure success.

Organization

The learner ordinarily has some *tendency to see and organize patterns or relationships* in the material or the activities which he is learning. But, *unaided, this tendency may not always form the desired relationship.* This natural organizing process must be given direction for effective learning to take place. Obviously, organization depends a great deal on *meaningfulness.*

The more meaningful the material is to the learner, the better he can organize and learn it. Meaningfulness generally is enhanced by a *preliminary overview* or introduction of the whole pattern of the material or process to be learned. It is best to use *words familiar to the learner,* and to explain all unfamiliar and technical terms as they are used. This is best done by relating new material and new terms to the *learner's past experience* by familiar illustrations and analogies.

Processes and procedures can be shown more meaningfully if the related equipment or mechanism is realistically simulated. However, the simulation should only be as faithful and as detailed as needed to convey the important factors he will encounter. *Simulation of procedures and equipment should emphasize only the essentials of the task* being learned at the moment and not overwhelm the student with unimportant details. Overly faithful simulation can detract from learning.

Generally, explanation of why things are done does help establish meaning. Here again, however, *too much detail too early can be harmful.* It is better to give only rather superficial explanations while a procedure is first being learned, and to introduce more detailed reasoning only after the procedure is fairly well established. The same can be said about explanations of how the equipment or mechanism works. Too much detail too early in the learning process is detrimental here also. Generally, it is best to give only the minimum information needed to learn the procedure. Detailed information on *"how" and "why" should come after the procedure is fairly well established.*

Finally, meaning can be strengthened further by reviews or summaries for logical units of material during each period of presentation, and at the end of the presentation. *Always tell them what you've told them.*

Organization and learning are easier when the material is grouped or *organized into patterns.* There are several kinds of patterning, all closely interrelated, that are useful for organizing material. Objects and ideas can be *functionally patterned.* For instance, the interconnected parts of an electrical system—a battery, a fuse, the wire, a switch, and a bulb—form a complete functional unit. An example of functionally interrelated ideas would be the concepts of current, voltage, and resistance. *Spatial patterning* is sometimes helpful—for instance, the grouping of the instruments on a control panel by function; or the interrelation of the parts of a planetary gear system. *Temporal* patterns are also useful in organizing material—for instance, the sequence of events in a complex procedure such as the disassembly of a carburetor; or the steps involved in solving a certain mathematical problem. Often, the procedure or material being learned falls into *a logical pattern* because of the combined elements of functional, spatial, and temporal organization. An example of this is trouble shooting procedures for isolating a malfunction in a television set; or, solving a word problem in algebra.

Participation

The individual learns only by his own activity—mental and physical. What he learns are *the mental and physical responses he makes and organizes himself.* Within certain limits, the more active the mental or physical learning behavior, the greater the learning. Mental participation is enhanced if *straight information is interspersed with questions.* Questioning can be effective, even though there is no expectation of the group or an individual answering. When questions are used in this manner, they should be followed by a slight pause and then answered for the students. Interspersed questions are even more effective *if the learner is required to answer* not just to himself, but by actually writing or reciting his response.

Learning a skill or procedure from a demonstration (motion picture, television, or a live performance) is enhanced when the learner has an opportunity to actually *practice the procedure during pauses in the demonstration.* The amount of participation and involvement of the learner varies considerably with the kind of practice. The learner can merely *visualize the task* or idea to himself, mentally picturing the steps involved. Or, going a little further, he can *imagine the feel of the task*, as if going through all the motions. A further step in participation would be for him to *verbalize the task or idea,* to formulate or to restate the verbal descriptions of the task as well as visualizing and imagining each step. Of course, the most effective of all would be to combine all of the above with *actual practice of the task itself.* Step-by-step demonstration films, slide-tape presentations, or demonstration lectures interspersed with the opportunity to practice the actual procedure or task are highly effective. A further extension of this process is participation by role-playing. Having the learner act out an assigned role in a situation can be a fruitful learning experience.

Confirmation

Much of the time in learning there is a selection process by which the individual tends to acquire and repeat those actions which did one or more of the following:

- Led to success in that situation.
- Tended to satisfy the motivating conditions.
- Served as a means to desired ends.

Unsuccessful or annoying actions tend to be avoided and to shift the learner's activity to other actions that may lead to success.

If there is to be full reinforcement of a response or action, there *must be a checking or confirmation* of the successes in the mind of the learner. He has to know whether the response or the action was right or wrong. More importantly, he must know *why it was right or wrong.* Once the right response is at least partially established, and provided the consequences of the wrong response are stressed, *demonstration of the wrong response* will further establish the right action.

Repetition

Mere repetition of an activity has very little, if any, strengthening effect upon learning. Repetition apparently just gives the strengthening or weakening factors more time to affect the learning. *Practice is effective only when done under favorable conditions,* such as proper motivation, meaningful organization, continuous participation, confirmation, etc. Purposeless or meaningless repetition and practice do not enhance learning. In fact, practice carried out under improper conditions can preclude any further learning and may actually prove to be retrogressive in effect.

Some degree of *over-learning is very important* to learning. Retention is favored if, in the initial learning, the material is practiced or repeated beyond the point of its being barely learned. However, repetition or continuation of practice, to the point where serious fatigue or boredom sets in, may have a negative effect upon learning. Variations, in the less essential details, tend to off-set the fatigue and boredom sometimes associated with over-learning and repetition. *Realistic practice* helps to overcome the demotivating effects of repetition. It is usually best to practice an activity in a realistic setting and in the way in which it will be used or needed in the future. Realism is most effective when only the essentials of the situation and the task are reproduced so the learner does not get lost in the nonessential details.

Generally, it is best to develop a task or a procedure, or even an idea, as a *whole-part-whole learning sequence.* For instance, when teaching the assembly of a carburetor, it probably would be best to begin by demonstrating the entire assembly procedure once; then to break it into a step-by-step demonstration interspersed with step-by-step participation; and finally to recombine all the elements in a complete run-through of the assembly procedure. However, the application of the whole-part-whole concept depends upon the degree of integration difficulty, the length of the material, and upon the ability of the learner. If a task is too difficult, or too long to be learned efficiently as a unit, it may be desirable to break it into meaningful smaller units first—but always emphasizing the relationships to the whole task. In this situation, meaningful sub-goals must be set for each sub-unit. Finally, the entire operation can be practiced as an entity by relating back to each of the sub-units.

Under many conditions, *learning is favored when practice periods are spaced* over intervals of time, rather than massed together. Concentrated practice of any one activity for periods much longer than thirty minutes without some sort of a break is open to serious question. A short break in the routine is usually all that is necessary. The change of activity is what is important, not the length of the break.

The sooner the practice follows the instruction or demonstration, the greater the benefit to learning. *Minimum delay of practice is the key to efficient learning.* Immediate imitation of movements while viewing them would be desirable, if such activity did not distract from the observation of the movements being demonstrated. Thus, intervals during the demonstration should be provided for practice. *Step-by-step intermingling of demonstration and practice* is usually the best way to learn complex procedures. However, provision must be made for *well-integrated review* and practice of the material or procedure in its entirety, pulling together all the discrete steps and activities that go to make up the complete procedure.

Application

The tendency or ability of the learner to generalize what is learned and to apply or transfer the new knowledge to new problems and situations is helped by *inducing a positive set or attitude towards generalization* as part of the instruction. Application and transfer to a new situation cannot be assumed—*there must be specific training for transfer.* Practice must be in a realistic situation to begin with; but, in addition, transfer of the new knowledge or skill to new situations or applications must be demonstrated and practiced by the learner. The learner must be made aware of the possibilities and problems involved in generalizing from previous knowledges and skills. Generalizing and applying newly learned material are facilitated by the *recognized similarity* of what has been learned to the actual task situation. Instruction should point out and foster the recognition of such similarities.

Individual Differences

Up until this point, the discussion has centered upon conditions for learning that can be manipulated and controlled in the learning situation. But what about those factors which the learner brings to the situation? *Individual differences among the learners are probably as important as all the other factors combined.* A person's ability to learn is greatly affected by his:

- General mental ability (intelligence).
- Educational level.
- Previous knowledge or skill in relationship to specific material to be learned.
- Specific aptitudes; for example: mechanical, spatial, verbal, perceptual speed, kinesthetic sensitivity, etc.
- Facility for learning through the eye or the ear.
- Attitudes or interests.
- Past experiences with various teaching techniques, such as film, television, and lectures.

These individual differences must be recognized and dealt with because of their considerable effect on all the other factors. To be truly effective, *instruction must be responsive to individual differences.*

Unfortunately, we today know very little about compensating for specific individual differences. Attempts to match "learning styles" with various instructional methods and media have, for the most part, produced rather inconclusive results, so far. The state of the art at this time is such that we are probably *wasting our energies and resources trying to devise instruction exactly suited to the learning needs of each individual student.* Perhaps future research will succeed in defining the qualities and kinds of instruction needed by the various types of learners. But, even if today we had that knowledge, there would still be a question of whether the schools could afford the developmental and operational costs of achieving that ideal. Moreover, there are strategies at hand now that present practical alternatives and closely approach the ideal of completely individualized learning.

Recent research and reinterpretation of earlier findings have led many educators to the view that *aptitude can be thought of as the amount of time needed by a learner to attain mastery of a learning task.* Implicit in this viewpoint is the belief that, theoretically, *all* students can achieve mastery of a particular learning task if each is given as much time as needed. In fact, it has been shown that *the effects of individual differences tend to disappear where students can learn at their own rate.* A clear demonstration of this proposition is the fact that, on standardized achievement tests, the top scores achieved by only a few students at one grade level are achieved by the majority of the students at a later grade level.

Although learning rate may well be the single most effective agent for compensating for individual differences, the sheer amount of time spent in learning does not account entirely for the level of learning attained. Each student should be allowed the time needed to master a learning task, but *the amount of time needed will be affected by both the individual differences and the quality of the instruction.*

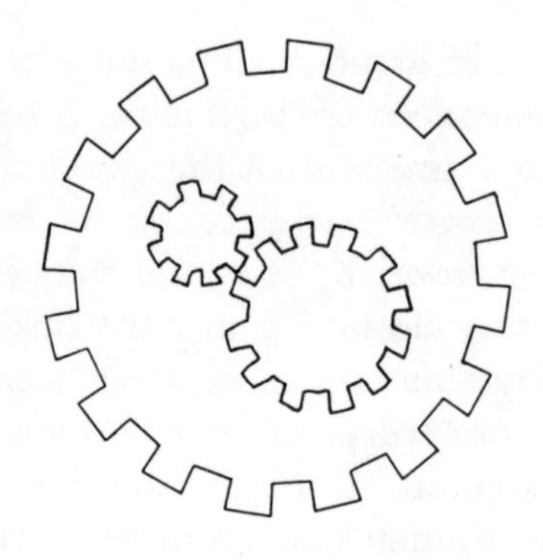

CHAPTER III

KINDS OF LEARNING

Introduction

The preceding chapter reviewed the general conditions that usually facilitate most kinds of learning. However, good instructional systems design calls for careful and selective application of these general rules for learning, because the *different kinds of learning need different conditions.* There are several identifiable categories of learning which, independent of subject matter, need a particular set of facilitating conditions. This chapter will describe these different kinds of learning and identify the special conditions for each.

Learning is said to have taken place when the learner can demonstrate a change in performance capability—*when he can do something that he could not do before.* There seem to be several distinctive kinds of performance that probably result from distinctive kinds of learning. The optimum conditions for each of these basic types of learning differ markedly. For instance, the best way to learn to set the timing on an engine would be to actually practice the adjustment of the distributor while using a timing light to check on the number of degrees of advance. This kind of procedure could hardly be learned by merely listening to a lecture or by only reading a manual. On the other hand, a student could not be expected to learn the theory of distributor operation by practicing the adjustment of the timing, no matter how many times he practiced the procedure. An illustrated lecture or text would certainly be better for learning about the theory of operation. In contrast, good work habits and attitudes cannot be instilled in a student by classroom lectures about the importance of being prompt, courteous, neat, etc. He is much more apt to develop these desirable traits if the entire system of living, training, and education clearly demands and openly rewards such behavior. Role-playing and case study exercises based on the students' real-life experiences would probably further reinforce the desirable behavior. These contrasting examples should serve to illustrate how markedly the different types of learning affect instructional design.

The following descriptions of the different types of learning generally follow the models detailed by Robert M. Gagné in his influential book, *The Conditions of Learning* (1965). The kinds of learning outlined by Gagné are particularly relevant to occupational education because *each type is based on, and described in terms of, a specific performance capability.* Moreover, the categories have practical application for analyzing the instructional content of a vocational training program because, regardless of subject matter, training objectives will logically fall within these performance capabilities.

Seven categories of learning are identified and listed in order of the complexity of behavior, from simple responding to highly complicated problem-solving. However, there really is no exact point at which one type leaves off and the next begins. Actually, what is described is *a hierarchy of behaviors in which each type leads into and supports the next higher kind of performance.*

Specific Responding

This is the simple, unitary, isolated act of making a specific response to a specific stimulus. The learner acquires *a precise response to a precise stimulus.* What is learned is a single, simple connection, association, or identification; for example, learning the names of new objects or items; or learning how to hold a new tool; learning to say "rivet" when shown one; or learning how to grip a glass cutter correctly. However, *the response is almost completely physical in content.* The individual has learned to respond only to the physical sight, sound, and feel of the word or the object, not to the ideas. What a rivet or a glass cutter does is not included in this simplest kind of learning. That knowledge falls within the higher orders of learning categories.

Even though the typical young adult has acquired a tremendous repertoire of this simplest kind of learning, it would be dangerous to assume too much; for many students often have conspicuous and unpredictable gaps in their backgrounds.

Motor Chaining

The simple, specific motor responses described above usually become part of a longer chain, such as in printing or drawing. What is learned is a chain of two or more specific motor responses, each of which is linked to each subsequent response. Motor chaining is the capacity for *carrying out a sequence of linked motor responses leading to a unitary act or product.* A very simple example would be the act of drawing the symbol for a resistor. Using a socket wrench to remove a spark plug is an example of a slightly more complex motor chain. Though both of these actions are composed of a fairly intricate chain of specific motor responses, they both are basically very simple activities.

Verbal Chaining

A verbal chain is essentially *a sequence or list of words linked together by associations of some sort.* The between-word associations may be real and logical or they may be artificial prompts created by or for the learner to help him forge the chain. Whether real or artificial, the linking associations are little used once the chain is well established. Telephone numbers, common phrases, and even long passages of prose or poetry can be recalled in their entirety after apparently having been forgotten for years. Recalling certain verbal chains—such as rules-of-thumb, lists of ingredients, numerical values, and the steps in a process—is often very important to job performance.

The ability to carry out complex procedures, such as the assembly of a carburetor, is based on a combination of motor chaining and verbal chaining. The recall of an operating procedure is dependent upon a chain of responses linked together by both verbal and motor prompts or cues. It is very common to hear a student talking himself through a procedure; and even after a procedure is well learned, there is usually some degree of internal vocalization.

Discriminating

To discriminate, the individual must learn *to make different identifying responses to two or more different stimuli.* In other words, the various aspects of the environment must be distinguished from each other whether they are colors, shapes, sizes, sounds, or even ideas. Discriminating is a relatively simple task when there are gross, self-evident differences. However, when the objects or ideas closely resemble each other, discrimination becomes more difficult. A student can easily learn to discriminate between a hammer, a bolt, and a wrench; but he also has to learn the more difficult discriminations between a ball-peen hammer, a tack hammer, and a carpenter's hammer—or the even more difficult discrimination between machine bolts of differing size and thread pitch. Because of the lack of easily distinguishable differences, the most difficult of all discriminations are among the various aspects of the social environment. Correct identifying responses to the different social stimuli are difficult to learn unless the consequences of the response are clearly foreseen; and, unfortunately, that is not often the case.

Classifying

A great deal of learning is concept learning. Basically, concept learning is acquiring *the capability to classify objects or events into categories.* To do this, the learner must acquire the capability of making a common response to a class of events or objects that may differ widely from each other in appearance but still have at least one important common feature. That is, he is able to make a response that identifies an entire class of objects or events by assigning them to classes of like function. Concepts are used to classify concrete objects, such as tools, into functional categories. For instance, a claw hammer, a tack hammer, a ball-peen hammer, and a blacksmith's hammer would all come under the concept of hammer; for, although their physical appearances differ widely, they are all tools intended for pounding. Though they differ in detail, they each share the overriding common characteristics of having a head and a handle, and being used for pounding. There are, of course, much more abstract concepts, such as empty, heat, electrical charge, liquid, reliability, and integrity. The more abstract and complex concepts are apparently acquired through verbal imagery of words that refer back to simpler concepts.

Rule-Using

Rules or principles form a large portion of the systematic knowledge that must be acquired during education of any sort. *The individual has to learn to perform tasks or activities according to certain rules or principles.* In simplest terms, a principle is a chain of two or more concepts that are linked in a cause and effect relationship by stating, "If A, then B," where A and B are both concepts. For example, when the two concepts, liquids and boiling, are linked in that manner they become a principle stating, "If it is a liquid, it will boil," or simply, "Liquids boil." A lot of time has to be spent in every training program learning rules—learning how to apply the principles, not just learning to state them. *General principles must be recast as specific behavior-controlling rules to be usable in training or on the job;* for instance, the general principle, "atmospheric pressure varies inversely with altitude," has little meaning for a vocational student until it is restated in terms that apply directly to the subject matter. Thus, the statement, "The fuel-air mix in an internal combustion engine becomes richer as the operating altitude is increased," is more meaningful; but even though the principle has been rephrased in subject-oriented terms, it is still too general to have any reality or utility. Only when the principle has been narrowed down to a performance-oriented rule such as, "Always readjust the carburetor for a leaner mixture when an engine is to be operated at a higher altitude," can it be used to control behavior. The same general principle concerning atmospheric pressure can be narrowed down to other specific rules in entirely different subject areas; for example, "Use a little less baking powder or a little more flour to adapt a regular recipe to higher altitudes."

Problem-Solving

Basically, the activity of problem-solving is *the solving of a novel problem* (a problem to which no known rule apparently applies) *by combining two or more previously learned rules or principles which leads to the creation of a higher-order rule or principle.* It is this new higher-order principle which is applied and is then learned. Of course, the new problem and new solution to that problem may be new only to that individual. What emerges from problem-solving, or "thinking" as it is commonly called, is a new capability in the individual—the ability to solve that kind of problem by applying the newly created principle. The next time he encounters a similar problem he probably will no longer consider it a problem, because he has learned to cope with it.

Examination of a simple problem-solving situation will illustrate the process. A student finds he cannot loosen a bolt. No matter how hard he tugs on the wrench, the bolt will not budge. This is something that has never happened to him before, although it is a common experience to most who have done any car repair. Problem: how to loosen that bolt? He realizes he somehow has to increase the amount of force he can apply to the bolt. He knows a wrench is a kind of lever (Rule No. 1), and he also knows that the longer the lever, the more the force exerted (Rule No. 2). Suddenly, he puts the two together and "discovers" the new rule that a longer handled wrench will probably solve his problem. Or, he might use entirely different rules to arrive at an entirely different solution. For instance, he might realize he must somehow make the bolt easier to turn. He knows lubrication lessens friction (Rule No. 3), and he knows oil is a lubricant (Rule No. 4), and he knows that a light-weight oil will flow into tight cracks (Rule No. 5). So, he puts these three together and "discovers" the idea that oiling the bolt will probably make it easier to turn. He might also come up with an even higher-order rule by combining his two newly created rules—"Whenever a bolt won't turn, soak it with light oil and get a longer handled wrench."

Although there appears to be considerable overlap between problem-solving and the simpler kind of learning involved in rule-using, there is a key difference. In rule-using, the individual is given a rule (solution to the problem) and must learn to apply it first in specific situations and then generalize to similar problems with it. However, *in problem-solving, the individual is not given nor does he have a ready-made rule or solution which he can apply to the problem.* Rather, he must discover for himself a new solution (new to him at any rate) by combining previously learned rules, with little or only indirect guidance from others.

Obviously, there isn't a ready-made solution for each problem that comes up on the job, nor are there specific rules of social conduct for every situation faced by the individual. Even if there were such ready-made solutions, it would hardly be practical to try to teach them all. Of course, each body of knowledge has an undergirding core of concepts and principles that have wide application. In addition, there are problem-solving techniques and strategies that can guide the thinking process, and these can be learned. Systematic logic analysis and the scientific method are examples of strategies that can be applied to many kinds of problems regardless of the subject matter content. *Strategies for systematically attacking problems can be learned.* For instance, using the split-half technique for locating a malfunction in an electrical circuit is much more efficient than randomly checking on a trial-and-error basis all the components in a radio or the electrical system of a car.

Learning Structure

Each of the seven types of learning described in the preceding section *establishes a different capability in the learner;* but, in addition, *each begins with and is dependent upon a different previously learned capability.* The learning of principles usually calls for the prior learning of subordinate principles; and these are built in turn on prerequisite concepts, discriminations, motor and verbal chains, and specific responses. The sequence of learning progresses from the simple to the complex, building on each of the intermediate steps. The hierarchy of events is as follows:

1. The learning of *specific responses and associations* is prerequisite to the learning of

2. *verbal and motor chains,* which are prerequisite to the learning of

3. *discriminations,* which are prerequisite to the learning of

4. *concepts,* which are prerequisite to the learning of

5. *principles,* which are prerequisite to the learning of

6. *higher-order principles* such as strategies for problem-solving.

The optimum set of conditions for learning is different for each capability, and one of the important differences is in the capabilities that the learner must have previously acquired and retained. If learning at any level is to take place with any degree of facility, all of the more fundamental capabilities must be present at the starting point. The problem of loosening the stubborn bolt could never have been solved without the student having previously learned the principles for lubricating with oil and for applying leverage. It also follows that those principles could hardly have been acquired before he learned the concepts of a lever, a wrench, and oil. Prior to that, he would have to be able to discriminate between a hammer and a wrench, light oil and water, and a short lever and a long lever. Probably, before that he would, by forming chains of motor and verbal responses, have learned the procedures for loosening a bolt with a wrench and for lifting a heavy object with a long bar. And, at some time or other, he must have initially learned to associate the word "wrench" with a certain metal object, the word "oil" with a slippery feeling substance, and how to grasp a bar in preparing to pry with it.

It should be pointed out that although learning usually follows a sequence of events somewhat similar to that of the oversimplified example above, *there can be alternate routes to the same capability.* In adult learning, especially, there can be considerable overlapping and combining of events, and even reversals of sequence. Even for adults, however, it is probably better in most cases for the learning events to occur in the right sequence, always building from the simpler to the more complex behaviors. Acquiring even quite complex capabilities can be relatively simple and rapid if all the prerequisite skills and knowledge are already in the learner's repertoire. Conversely, *it cannot be overemphasized that learning becomes difficult and inefficient if any one of the intervening steps is left out completely.* All the intermediate capabilities are needed to facilitate learning–failure to develop even one will adversely affect the process.

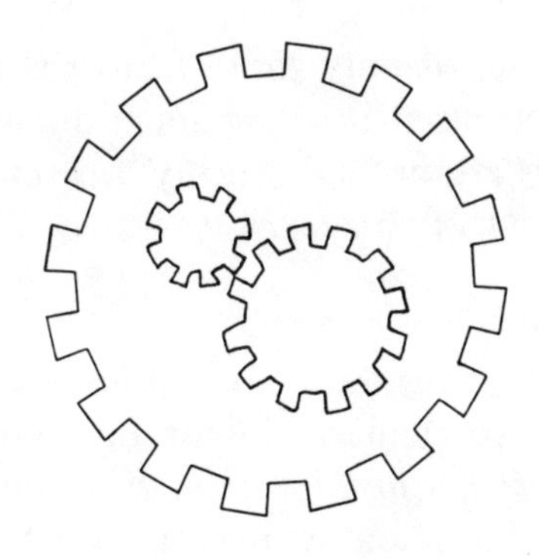

CHAPTER IV

FACILITATING LEARNING

Introduction

Throughout the preceding discussion, it has been emphasized that good instructional design demands selective—not across the board—application of the principles of learning, because different kinds of learning require different sets of conditions. *The factors that influence learning—MOTIVATION, ORGANIZATION, PARTICIPATION, CONFIRMATION, REPETITION, AND APPLICATION—must be applied according to the special needs of the different kinds of learning.* For some kinds of learning, primary emphasis has to be placed on two or three of the principles while others are de-emphasized; and, for other kinds of learning, the emphasis must be shifted completely. For instance, repetition is important for establishing the motor chains involved in freehand lettering. On the other hand, one-trial learning is a very common outcome of problem-solving. Once an individual has discovered a new rule or principle, he can usually remember it later without having to practice the discovery. No one is likely to forget how to get a balky radio going once he has discovered that a sharp rap on the cabinet will start it playing again.

The special requirements of each kind of learning are discussed in the sections that follow. For the sake of brevity, mention will be made of only those factors that are of particular significance to the type of learning under discussion. Even though the other principles are not specifically mentioned, they *all have a part to play in facilitating the learning.*

Specific Responding

Making a precise response to a precise stimulus. The three factors that play a particularly important role in learning simple associations are *Participation, Confirmation,* and *Repetition.* Specifically, the associations and connections are made by repeated exposure to the response-provoking stimuli with immediate confirmation of the active response. Repeated vocal or written responding is usually needed to learn new technical terms. Merely looking at the list of words is not enough. Likewise, a student would have to practice actually gripping a Leroy lettering pen to get the feel of the act itself. He probably could not develop the right grip by just watching someone else or by having the grip described to him.

Motor Chaining

Carrying out a sequence of linked motor responses leading to a unitary act or product. The special conditions for acquiring motor chains are essentially the same as those for specific responding—namely, *Participation, Confirmation,* and *Repetition.* However, the purpose is to link a series of motor responses, not to establish individual connections. With the help of external prompting (directions), the separate motions are performed in close sequence. Repetition brings about progressively smoother performance by establishing a stimulus-response link between the successive motions. *Each move in turn comes to function first as confirmation for the move that precedes it, and then as the stimulus for the move that follows it.*

If the first motion is incorrect or incomplete, then the second cannot possibly carry on from there toward successful completion of the total chain. However, if the first move is correct and complete, then the second can take over from there. Thus, being able to proceed successfully with the second motion is confirmation of the first. For example, a key has to be pushed all the way into the lock before it can be turned; but once it can be turned, the correctness of the first motion (pushing key all the way in) is confirmed.

Once the second motion is complete, having already served to confirm the first, it then becomes the stimulus or cue for the third motion in the chain. The third move then confirms the second and cues the fourth, and so on, until the entire action is completed. In this way, each separate move actually becomes the mediating link between the moves that come before and after it. After these links have been forged, the prompting from demonstration and verbal directions is no longer needed.

Here again, it would be highly unlikely that a student could learn to letter with a Leroy pen by just watching someone demonstrate or by having the motions described. Although demonstration and verbal prompting are important initially, he has to develop the feel for the right motions through practice if he is to become proficient.

Verbal Chaining

Producing a sequence or list of words linked together by association of some sort. The conditions for learning verbal chains are very similar to those of motor chains—*Participation, Confirmation,* and *Repetition.* A series of external prompts or cues are used initially to evoke the verbal responses to be chained, but with repetition *the words themselves come to serve as confirmation of the preceding words and prompts for the succeeding words.* Eventually, the external cues are no longer needed, and the chain becomes self-starting and self-sustaining. The stimulation provided by hearing and feeling one's self say the first word becomes the cue for saying the second word; the saying of the second word cues the third, etc. The same sensory cues produced as the second word is spoken probably also serve to confirm the correctness of the first response.

Before the words themselves become the mediating links, *the chain is put together and practiced with the help of outside prompts of some sort.* In some instances, the individual may create his own prompting devices, such as associating a man's name, Alfred Lyons, with "A. Lion" because of his unruly hair. Often, the prompting device is furnished the learner as an artificial aid to memory. In music, for instance, the word "face" prompts the recall of the notes F, A, C, and E that fall between the lines of the staff; while the sentence, "Every Good Boy Deserves Fun," prompts the recall of the notes E, G, B, D, and F that fall on the lines. The easily remembered common word and simple sentence stimulate the recall of the hard-to-remember chains of letters. The artificial prompting devices usually lose their importance as the chain is established independently. Once well established, verbal chains, and sometimes the prompting devices as well, are often recalled years later in their entirety.

In some cases, there are no external cues, and the chain is established through sheer repetition. The value of π, 3.1416, is an example of a verbal chain held together purely on the strength of the stimulation provided by seeing, saying, hearing, and feeling each of the succeeding words—there are no external cues, only internal. *Obviously, Participation, Confirmation, and Repetition are especially important to the learning of verbal chains.*

When motor and verbal chains are combined in the form of complex procedures, such as the alignment of the three stages in a superheterodyne radio or putting a carburetor together, the same sort of externally prompted practice is needed to establish the chains initially. Demonstration, discussion, and illustrated manuals giving step-by-step directions can all provide the kind of prompting needed at first. Eventually, the motor chains, along with internally vocalized chains of verbal prompts, become the mediating links between each of the succeeding steps involved—acting to confirm the previous step and to cue the follow-on step.

Discriminating

Making different identifying responses to two or more stimuli. To discriminate among various objects or events, the different stimuli have to become associated with different responses. This means that, in addition to learning what an object is, *the individual also must learn what it is not.* Therefore, there must be *practice providing contrast of the correct and incorrect stimuli.* The stimuli must be presented in a manner that emphasizes their distinctiveness. In some ways, the special requirements are quite similar to those of specific responding: *Participation, Confirmation,* and *Repetition*, but *Organization* is an added requirement.

The individual frequently must discriminate to some degree in nearly all kinds of learning. Most learning involves verbal and motor chains, and an important part of chaining is discriminating among the various stimuli and responses that make up the chains. As one would expect, there is usually considerable interference among all the different stimuli and responses, leading to both the confusing and the forgetting of the chains previously learned, the chains being learned, and the chains subsequently learned. *To prevent interference, the distinctiveness of each member of the set to be discriminated must be emphasized by providing contrast practice* in which the various combinations of stimuli requiring different responses are contrasted with each other. For instance, to learn to discriminate among the different kinds of bolts, the student should probably lay out samples of machine bolts, stove bolts, hexheads, roundheads, etc., where he can directly contrast and compare them, thus learning what each is not, as well as what each is.

Classifying

Making a response that identifies and categorizes an entire class of objects or events. If the prerequisite individual identifications, motor and verbal chains, and discriminations have previously been made, it is relatively easy for the adult to learn the relevant concept. A single exposure to a number of examples that cover the range of differences within the class, while the important common features are verbally emphasized for him, usually brings about concept learning.

Organization, Confirmation, and *Application are particularly important to concept learning.* The exposure to the examples that represent the class being learned needs *direct verbal guidance to organize the relationships and similarities* among objects or events with wide ranges of apparent differences. Representative samples may include the actual objects, actual experiences, pictures that illustrate the objects or experiences, verbal portrayal of the objects or events, or possibly a combination of all of these. Although concepts can sometimes be grasped entirely through verbal communication of the idea, it is best to *furnish concrete applications of the concepts wherever possible.* Often, if only words have been used to describe the class of objects or events, only the words are learned; not a concept with real application. *Confirmation of the acquisition of the concept is an integral part of the learning process,* as well as a means to verify whether the concept has been grasped.

Typically, learning to identify a class of objects to establish a concept follows a process something like the following example. Suppose a student does not know what a rivet is; he does not have the concept. He might, first of all, learn just to say the word itself in response to seeing a sample rivet. Next, he would probably learn to discriminate a rivet from somewhat similar objects such as bolts and nails by using verbal chains to state, "That is a rivet. That is not a rivet. That is a bolt," etc. At that point, he can say the word, associate the word with an actual rivet, and discriminate a rivet from a nail or a bolt. Then, as the first step in acquiring the concept itself, he would compare several kinds and sizes of rivets, noting that despite obvious differences in appearance, they all have the same general characteristics: no thread, no point, a broad head, short in length, and apparently made of soft metal. Now all the prerequisite learning has taken place. He can correctly respond, "That is a rivet," when shown each of three different kinds of rivets among some nails and bolts. What might be called a "tentative" concept has been formed at this stage. The final step in concept learning is taken when he has correctly identified a fourth kind of rivet he has not previously seen. With confirmation of the correctness of this response, the concept is learned–the tentatively formed concept has been put to the test and found to be adequate. *A concept is learned only after a trial application has confirmed its adequacy.*

Once the concept has been confirmed, and if the range of samples has been broad enough, the student should then be able to generalize to any new type of rivet and to classify it as such. There probably is *not much need for repetition in concept learning if all of the conditions are optimal.* If repetition is needed, it is probably because all of the prerequisite capabilities were not present initially, or one of the intervening steps was slighted.

Rule-Using

Performing tasks or activities according to certain rules or principles. A great deal of a person's lifetime of mental and physical activity is spent acquiring and using both concepts and principles. With so much time devoted to learning concepts and principles, both in formal schooling and in day-to-day living, it is fortunate that the conditions for learning—except for the prerequisites—are very nearly the same. *If the right prerequisites are present for each, the learning of both concepts and principles can take place under similar conditions and by nearly the same process; Organization, Confirmation, and Application are of primary importance to both.*

Principles and rules are basically formed by combining two or more concepts in a meaningful relationship. Therefore, when a principle is to be learned, the individual must first know and understand the concepts to be linked. Verbal instructions and guidance, along with demonstration of the ideas involved, are used first to recall the relevant concepts and then to organize the relationships for the learner. The learning process *should be structured enough to insure successful "discovery" and application of the rule.*

The first step in the prescribed sequence of events is a *verbal description of the kind of performance that will demonstrate successful learning of the rule.* This information might be given in the form of a question such as: "What simple adjustment will probably make an engine run better at high altitude?" The learner has not been given the rule, but he has been given the expected outcome of applying the rule—the engine will run better at high altitudes.

The next step in the process involves leading the student to recall the relevant concepts and sub-principles that are components of the rule. *Directive questioning, and possibly demonstration, should be used to prompt the recall of the needed concepts and sub-principles;* for example: "What happens to the air as you go higher? Right, the air gets thinner, and thus less oxygen. So, if the air gets thinner, what happens to the fuel-air mixture? That's right, the mixture gets richer. And, why does it get richer? There is less oxygen in proportion to the fuel." Etc. At that point, the relevant concepts and sub-principles have been recalled—air gets thinner at high altitudes, and thinner air makes the fuel-air mixture too rich.

The learner is now ready for the next step in which he is *led by verbal cues to formulate the principle for himself.* Again, a question might be used: "What should you do to correct the mixture for a high altitude operation?" The student could then respond by stating the rule in his own words: "You should lean out the mixture for high altitude operation." Thus, through a series of leading questions, he has been guided along to self-discovery of a tentative principle.

The stage has now been set for the final learning act in the process of principle acquisition—*the confirmation of the tentative rule by application.* In this step, the learner should be asked to demonstrate the effectiveness of the rule by citing one or more concrete instances of its application. In this case, he might state that the mixture can be leaned out by adjusting the carburetor. Or, he might demonstrate the application of the rule by actually readjusting a carburetor that is set for too rich a mixture. Either verbal or actual demonstration of the adequacy of the rule or principle is a very important part of the learning process. The additional step of formalizing the statement of the principle may be taken by having the learner produce a refined version of his tentative rule to *make sure it is complete and accurate.*

It is, of course, entirely possible to learn the more simple principles rather quickly without going through the entire series of events described above. *Adults can often learn a new principle entirely from the verbal statement of that principle.* However, there is always the danger of learning merely a verbal chain, with no real understanding of the principle involved, unless there is confirmation. Description of a situation where the rule applies may suffice as confirmation. Even for the adult, however, it is probably best to include some actual practice in applying the rule where possible.

Usually, the series of learning events leading to rule acquisition does not need to be repeated. Once a rule has been arrived at by this process, *the learning seems to be unaffected by repetition, and what was learned is highly resistant to forgetting.* Even though the learning of a rule often takes place on a single occasion, *practice may be needed to develop skill in using the rule.* Although a student may learn that the carburetor should be readjusted for high altitude operation, he will also need to practice adjusting the carburetor before becoming skilled at it. The same situation applies to mathematics: A student may have learned the rules for finding the least common denominator, but he will have to solve a series of practice problems before he becomes skilled in using the rule.

As shown in the example of rule-learning above, principles are seldom learned in isolation. Instead, a student generally learns sets of related principles that form the organized knowledge in a certain subject. *Organized knowledge is basically a hierarchy of principles.* Two or more concepts may be prerequisite to the learning of a principle; for example: air, higher, thinner, and altitude must be understood before putting together the principle: air gets thinner at higher altitudes. Similarly, two or more principles may be prerequisite, thus subordinate, to a higher-order principle. Two sub-principles form the third and higher order in the example above. Of course, this higher principle might be a part of an even higher principle, and so on. Thus, the entire set of principles forms a hierarchy that is the organized knowledge of a topic area.

Problem-Solving

Solving a novel problem by combining previously learned rules to create a higher-order rule. As pointed out earlier in this chapter, problem-solving is more than just combining two or more previously learned principles to arrive at a solution to a problem that is new to the individual. More important is the fact that, *in the process of arriving at a solution, a higher-order principle or rule is discovered and learned.* The higher-order rule then becomes part of that individual's repertoire to be used again with similar problems.

Successful problem-solving always depends on the previous experience of the learner, or more specifically, on the availability of previously learned principles. *The special conditions and process required for problem-solving are very similar to those needed for rule-using.* In fact, the two capabilities are so similar that it is difficult at times to distinguish where rule-using leaves off and problem-solving begins. Even though *Organization, Confirmation,* and *Application* are of particular importance to both kinds of learning, there are some real differences, because problem-solving is basically a self-directed activity, with little or only indirect guidance from others. On the other hand, principle-learning usually involves considerable external verbal guidance.

Organization is very important in problem-solving; but the attack on the problem is usually organized without outside help. There is still the need for guided discovery in the form of verbal directions, but *the individual must furnish the guidance himself by using previously learned problem-solving strategies and techniques.* He may even have to figure out what the problem is in the first place, as well as to determine what kind of performance will indicate a successful solution. Of course, in a learning situation the problem and the expected terminal performance are usually stated for the student. However, in day-to-day living, both at home and at work, the individual has to attack many problems entirely on his own. He must organize his attempts to use various combinations of previously learned rules and principles. The better his attack is organized, the more efficient he will be at problem-solving. Therefore, previously learned strategies and techniques for problem-solving (which are rules in themselves) are very important prerequisites.

After the individual has arrived at a tentative solution, he must confirm the adequacy of the solution through trial application. Just as in rule-using, the final step in problem-solving is confirmation through successful application. *Only when the tentative solution has been tested and confirmed as adequate does the learning of the higher-order principle take place.* In problem-solving there is usually no need for confirmation from outside sources. The solution is usually self-confirming; if it works, it is adequate.

The length of time needed to arrive at a solution in a training situation is largely dependent upon the amount and kind of guidance provided. Unless there is some guidance, a student may spend an excessive amount of time learning a very simple principle. Often, when there has been no guidance at all, the individual will get off on the wrong track and never arrive at a solution. On the other hand, if only started in the right direction, he may be able to solve the problem. Thus, it is not always practical nor efficient to put an individual entirely on his own in a school situation. At the very minimum, there should be enough verbal guidance to get him started in the right direction. However, there is some evidence that *the more self-direction and self-discovery there is in the problem-solving process, the better the learning.* Moreover, the chances are greater that the solution will be remembered and applied to a wider range of similar problems in the future.

Learning Analysis

It is evident that for each of the seven kinds of learned performance there are seven different optimum sets of conditions. In the first place, each type of learning requires a previously learned capability on which to build, and ends up producing a different kind of performance; each begins and ends at a different level in the learning hierarchy. In addition, each type has a set of external conditions under the control of the instructional designer that determines the efficiency of the learning process and the retention and transfer of the learned capabilities. Some kinds of learning need a great deal of repetition, while others do not. With some, verbal communication and guidance are very necessary to the learning process, while with others they are of little or no help. Tables 1 and 2, at the end of the chapter, summarize the different kinds of learning and their special requirements.

Good instructional design demands careful analysis of the training course content to determine which kinds of learning will be involved at each stage in the individual's development. Detailed analysis is needed because any one hour of instruction may include any or all of the different types of learning and, hence, may require several different sets of facilitative conditions. General rules and methods for learning are of little help in designing instruction. The instructional method must be responsive to the specific kind of learning involved. Blanket application of the "lecture method," "demonstration method," "discussion method," "discovery method," etc., to an entire course or even to a single lesson is grossly inefficient and ineffective. Likewise, the rules of learning are not specific to subject matter—there are no "principles of mathematics learning" or "principles of auto-repair learning." *The conditions for learning are specific to the type of learning involved, not to the subject being taught.*

Table 1

Types of Learned Performance

Performance Type	Definition	Example	Inferred Capability
Specific responding	Making a specific response to a specific stimulus	Repeating "rivet" when shown one	Connection; Identification
Motor chaining	Exhibiting a chain of motor responses, each of which is linked to each subsequent response	Using a wrench to remove a sparkplug	Sequence of motions
Verbal chaining	Exhibiting a chain of verbal responses, each of which is linked to each subsequent word	Listing, from memory, the steps for starting a diesel engine	Verbal associations; Verbal sequence
Discriminating	Making different (chained) responses to two or more physically different stimuli	Pointing out and identifying a ball-peen hammer, a carpenter's hammer, and a tack hammer	Discrimination
Classifying	Assigning objects of different physical appearance to classes of like function	Sorting out all the resistors from a pile of spare parts	Concept
Rule-using	Performing an action in conformity with a rule which is composed of two or more concepts	Adding more flour for high altitude baking	Principle or rule
Problem-solving	Solving a novel problem by combining rules	Trouble shooting a radio	Principles, plus "problem-solving ability"

Table 2

Summary of Facilitative Conditions

Performance established by learning	Internal (learner) conditions	External Conditions
Specific responding	Certain learned and innate capabilities	Repeated exposure to response-provoking stimuli; immediate confirmation of active response
Motor chaining	Previously learned individual connections	Presented a sequence of external cues that call for a sequence of specific responses; repetition to achieve selection of response-produced stimuli
Verbal chaining	Previously learned individual connections and cues	Presenting a sequence of external verbal cues; effecting a sequence of verbal responses at the same time
Discriminating	Previously learned chains, motor or verbal	Practice providing contrast of correct and incorrect stimuli
Classifying	Previously learned multiple discriminations	Recalling discriminated response chain along with a variety of stimuli differing in appearance, but belonging to a single class; confirmed by successful application
Rule-using	Previously learned concepts	Using external cues, usually verbal, effecting the recall of previously learned concepts in a suitable relationship; confirmed by specific applications of the rule
Problem-solving	Previously learned rules and problem-solving strategies	Self-direction and selection of previously learned rules to effect a novel combination which is self-confirming

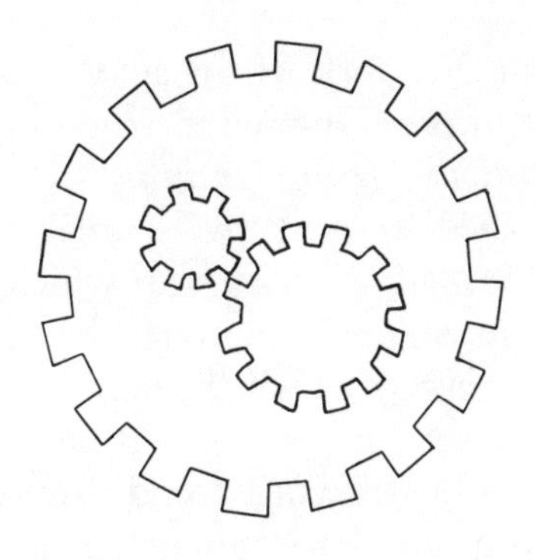

CHAPTER V

THE INSTRUCTIONAL SYSTEMS CONCEPT

Introduction

The earliest forms of training were highly individualized when knowledge and skills were passed on from father to son, from craftsman to apprentice, and from tutor to pupil. Individual tutoring reached its high point with the Socratic system, in which the student was led by a series of directive questions to self-discovery of knowledge. Everything else being equal, a one-to-one tutorial situation is still probably the best means of passing on knowledge and skill.

During the last 50 years, two powerful forces have brought tremendous change: the population explosion and the information explosion. Both the number of people demanding education and the amount of information demanded have been growing at an ever-increasing rate. Out of necessity, instruction of large groups by lecture and textbook was begun long ago. Recent instructional tools, such as slides, filmstrips, motion pictures, recordings, radio, and television can all be used with large groups of students. These new media are all commonly used today with varying degrees of success and acceptance and, when used correctly, they add a great deal to the learning process.

Recently, a combination of the Socratic method and learning

theory derived from the work of behavioral psychologists has led to the development of programmed instruction, a highly promising *technology for individualizing training.* In programming, the content to be learned is objectively analyzed, organized, and developed into an instructional system which takes into account learning theory, the subject, matter, the individuality of the students, and the means by which the content will be presented to the students.

Today there is a growing trend towards combining the techniques of group and individualized instruction to create instructional systems that are tailored to the specific needs of a particular learning situation. Training systems have been developed that combine television, film, individualized self-study, programmed materials, small-group instruction, and individual tutoring.

A TRAINING SYSTEM IS A SET OF INTERRELATED, INTERACTING, PRECISELY CONTROLLED LEARNING EXPERIENCES THAT ARE DESIGNED TO ACHIEVE A SPECIFIC SET OF TRAINING OBJECTIVES, BUT ORGANIZED INTO A UNIFIED, DYNAMIC WHOLE WHICH IS RESPONSIVE AND ADAPTIVE TO THE INDIVIDUAL STUDENT WHILE FULFILLING SPECIFIC JOB-RELEVANT TRAINING CRITERIA.

An instructional system is *both a product* (a complete instructional package) *and a process* (a method for designing the product). The instructional systems development process is a logical extension and combination of the principles of programmed instruction and the systems engineering concepts used in the development of complex defense and space systems. The application of good systems engineering principles along with those of programmed instruction across the entire range of media (not just the printed page) and to entire curricula (not just small portions) has introduced *empirical methods for the analysis, design, development, and evaluation* of education and training. The emerging instructional systems technology demands that all training course design decisions be based on analysis of student performance data. Each step in the development cycle must be empirically tested and validated against actual performance data. The training course developer now has a tool for determining the validity of training objectives, content, sequence, methods, media, and achievement. *The effectiveness and efficiency of training developed by this process can be described precisely by concrete student performance data.*

Throughout the remaining description of the instructional systems development process and product, the ideal process and the optimum product will be described. Realistically, there can be no such instructional system. System design and engineering are always a series of minor compromises in which one desirable characteristic is partially traded off for another. However, training systems development, like space systems development, must somehow arrive at the best possible mix of desirable features. The remainder of this book will describe the process for arriving at that best possible mix.

The Elements of an Instructional System

When the principles and methods of instructional systems development are applied to entire curricula and across the entire media spectrum, the result is an empirical process for the analysis, design, development, and evaluation of education and training. The key to the process is the *identification of the desired behaviors,* the specification of *what controls those behaviors,* and the determination of the *techniques needed to shape those behaviors.*

There are eight major attributes of an instructional system, each derived from parallel principles in programmed instruction and systems engineering concepts. The following sections will discuss each of them briefly.

Behavioral Analysis

A detailed task description and analysis is made of the tasks to be learned. The task may be to locate, remove, and replace a bad sparkplug in a car—a rather simple trouble shooting procedure. Or, the task might be to use a stock catalogue to get all the information and then order a special part for a carburetor—a rather complicated process requiring special skills in reading, indexing, writing, pricing, etc. The analysis describes and defines all the important skills and knowledge needed to perform the tasks. Specific training objectives are then identified for each task and sub-task. Each objective calls for a behavior that can be verified; an activity that can be measured accurately and reliably. By definition a performance objective describes precisely (a) what the student is to do as a direct result of his learning activity; (b) under what conditions and limitations; (c) to what level or standard of performance. The learning objectives demand *specific student responses* to *specific stimuli.* The more closely the performance objective matches the desired real-world behavior, the better the objective.

Optimum Step Size

The instructional content is organized into relatively small sequential steps, each requiring a desired *student response* (performance objective). Each step builds from the preceding response toward the succeeding step and its subsequent response. Only information and activity directly relevant to the desired behavioral outcome are included. Each step contains neither more nor less instruction than needed by the student to successfully perform the stated training objective.

Active Responding

The student has to interact with the instruction by responding in some specified manner at each step. The response may be replying to a question, performing a task, or a combination of both. The response and the stimulus for the response duplicate, as closely as possible, the demands of the real-world task and situation. When possible, the response calls for realistic application of the new knowledge or skill, not just a parroting of the facts nor the mere repetition of an action. By forcing the learner to make specific *overt responses* throughout the learning experience, his behavior is shaped step-by-step until he achieves mastery.

Immediate Confirmation

Immediately following each response, the student is given knowledge-of-results. He finds out at once if his response was right or wrong, and why. Making a correct response and having that *response confirmed* as correct strengthens the probability that he will respond appropriately in a similar situation later.

Managed Reinforcement

Learning is not left to chance. The student is firmly, but subtly, guided toward making the *correct response.* Thus, the training experience is a series of mostly successful experiences; generally giving positive, not negative, reinforcement. Even though the training situation is highly structured, it still challenges the learner by making him reason and work through the desired response and by allowing for some error.

Learner-Controlled Pacing

Normally, the students are not forced to progress at the same pace; however, they all must reach the same objectives and meet the same minimum criteria of success. The student is moved on to the next step only after he has succeeded with the preceding one; thus the learner's *responses control* the pace of instruction. Ideally, the majority of the training is individually paced, but, depending on media, content, facilities, resources, scheduling, etc., varying degrees of group-pacing may be imposed without seriously impairing the efficiency of the system. Even when training is completely self-paced, deadlines and mileposts are included to force progress to some degree.

Learner-Controlled Content

The content of the training system adapts itself to the student, not the student to the content. The learner gets only what he needs to know and only when he needs it. The information given the student is appropriate to his needs and abilities of the moment. If he does not learn, it is not his fault, but that of the system. Materials and tasks are organized to allow the student to pick his own best route toward the desired terminal behaviors. If need be, he is detoured from the main stream of instruction into some remedial tutoring material. On the other hand, he may by-pass material and interim steps any time he can demonstrate the desired terminal behavior. Thus, the content of instruction will vary considerably from student to student, because it is largely determined by each individual's *response patterns.*

Validation

The system is thoroughly tested to make sure it is capable of doing what it was designed to do. The objectives are validated against the real-world; the target population is tested to determine its entry-level skills and knowledge; content, sequence, method, and media are validated against student performance data; the system is tested bit-by-bit as each is produced, and then tested in its entirety after it is complete. The burden of success rests with the system, not with the student. If representatives of the target population cannot perform at the desired level of proficiency, the system is revised and retested and revised until they do. The analysis of the *response data* provides knowledge-of-results to the instructor, author, and curriculum developer.

Implications for Learning

There is nothing new or radical in these eight elements when taken alone. However, putting them together, and *applying them as unified concept, is new.* The concept is based on one key idea, a deceptively simple one: the systematic weaving of *student responses* into the fabric of instruction. Student response is the key. Each of the other elements derives from, and is dependent upon, student response. In fact, each of the above descriptions of each element contains that key word: *RESPONSE.*

When the eight principles of instructional systems development are applied to any media, *instruction becomes the two-way communication process it must be if learning is to take place.* Continuous and concurrent feedback (knowledge-of-results) is given to both the student and to the system (instructor, teaching machine, computer, curriculum developer, etc.), thus creating a closed-loop system. The principles can be applied to all means of communication, visual and aural. Moreover, it is a system of communication which potentially can insure near-perfect interchange of information (learning) because of the feedback to both ends of the loop.

A training system demands more than just the passive exchange of information, because feedback and knowledge-of-results include acting upon and reacting to the information. In effect, an instructional system is a self-regulating mechanism in which the information loops control the behavior of the student and the system (instructor, teaching machine, computer, curriculum developer, etc.). *Both the student and the system are in a cause-and-effect relationship to each other.* Each is dependent upon the other for information as to how well each is performing and, consequently, what each must do next.

Moreover, because of the enforced participation in the learning process, through performance of the desired behavior, each student is learning by doing. Thus, all media (books, slides, films, lectures, and television) are something more than just means of displaying or disseminating information. Instead, all media become means of directing and selectively controlling meaningful individual learning activity. The application of the principles of systems development to the different media is changing the concepts of those media. In effect, *an instructional system actually "individualizes" group instruction.*

The Development Process

There are essentially four phases to the instructional systems development process:

1. Specifying system objectives.
2. Developing the preliminary system design.
3. Developing, testing, and revising the system.
4. Installing and field testing the system.

There is a great deal of similarity between training systems development and the development cycle followed when designing and producing an automobile. A brief description of the steps involved in developing a new model car will illustrate the parallel processes. The first decisions to be made are concerned with establishing the development goals—the features the car is to have that are different from the old model. The demand for improvement and change, of course, comes from the customers. They want a better product. For instance, they may want a more powerful engine, more comfortable seats, and a new exterior design. However, before the customers' desires can be accepted as the development goals, there has to be a feasibility study. Is it going to be practical to develop what the customer wants? Is there enough demand for the changes? How much will the changes cost? How long will it take to make the changes? Will the customer be willing to pay the added cost?

Once the changes are judged feasible, the engineers can then set about drawing explicitly detailed objectives. Specific performance criteria have to be established, such as 300 brake-horsepower at 3500 R.P.M., engine operating temperature of 165 degrees, guaranteed engine life of five years, etc. The criteria established at the beginning of the development cycle will be used later to judge performance. Next, the preliminary design specifications are drawn up, and decisions are made as to how to make the engine more powerful, how to make the seats more comfortable, and what changes will be made in the external design. The engineers then draw up elaborate blue-prints that specify and describe every part and every detail of the entire system.

The next step is the actual development and testing of each of the component parts of the system. When the new and more powerful engine has been built, it is tested to see if it comes up to specifications, if it meets the objectives set for it. The seats are tried out to see if they are more comfortable, and the exterior design may even be tried out on potential customers. As each of the component parts of the entire system is produced, it is tested to see if it comes up to the performance specifications. If the engine does not perform as hoped, it is back to the drawing boards for revision of the design; then the engine is rebuilt and retested. So revising and retesting go on until the engine meets the criteria laid out earlier. The same thing may happen with the seat; it is tested and, if necessary, revised and retested. If sample customers do not like the exterior, it will be revised until it does satisfy.

Eventually, the complete car, with all its component parts, must be tested as a unit—the new parts with the old—because even components used in previous model cars may not work in combination with the new. So, the car is taken to the test track and driven under varying conditions, while careful note is made of its performance in relation to the objectives set out for it in the very beginning of the development cycle. Deficiencies that show up are taken care of by redesigning and reworking the deficient components. The final and perhaps most important test is given the new model when the customers use it on the highway. Thus, the final revisions of the new car are based on the customers' experience with it.

The development cycle for a training system closely parallels that described above. The instructional systems designer must go through almost the same sequence of events as the new car designer. Just as in new car design, the first step in the instructional design process is to *conduct a feasibility study.* However, in this case, the customers are the potential employers. What are the needs of the job market? What kinds of skills are needed? What does the employer want in a new employee? Is the training for certain jobs practical? Will it take too long? Will it be too expensive? These are the kinds of questions that must be answered in the feasibility study.

Once the decision has been made that a certain training course is desirable and feasible, the next step is to *conduct a task analysis* to determine the kinds of performance, the knowledge and skills, and standards of performance required by the job. Armed with this information, the systems designer can then *develop the training objectives*—that is, explicit statements of what the student must be able to do when he completes the training. Both terminal and interim objectives are needed to specify the behaviors, the conditions, and the standards of performance required to complete the training.

Next, the designer must *develop the criterion testing devices* that will be used initially to help determine the validity of the training objectives and used later to measure the effectiveness of the training system itself. Both the individual lessons and the entire system are pre-tested against the stated objectives. The criterion test is a performance test that calls for the application of the knowledge and skills learned in the course. The items in the test are derived directly from the objectives, with at least one item for each of the training objectives.

The next step is to *validate the criterion test* by giving it to trained and untrained sample populations to determine its reliability and validity. If the majority of the *untrained sample respond correctly* to a particular item, that item probably has no validity nor reliability. In contrast, if the majority of the *trained population do not respond correctly* to a particular item, then the validity and reliability of that item is suspect. The analysis of the test data is concerned primarily with perfecting the test as a measuring instrument.

These same test data are next used to *validate the training objectives* themselves. The same reasoning is applied when assessing the objectives. Where the *untrained population can respond correctly* to an item, there is probably no need to include its companion objective in the course. Similarly, if the *trained population cannot respond correctly* to a particular item on the test, then probably that objective is not needed in the course, because the trained worker can get along on the job without that knowledge or skill. Thus, by giving the criterion test to trained and untrained samples, the test itself is validated and the training objectives from which the criterion test items are derived are also validated. At this point, the initial design phase has been completed, and the objectives of the system have been clearly laid out.

Following the preliminary design stage, the detailed blueprints for the training course are laid out. First, the *learning structure and sequence are developed*—the building blocks of knowledge and skill are laid out in proper sequence; and then the course material and content are outlined accordingly. Once the content sequence is outlined, the next step is to *develop the learning strategies.* The special set of conditions for each kind of learning must be specified and the decisions on media, method, and materials must be made at this point. With the instructional strategies decided upon, the preplanning and detailed design phases are completed.

Using the blue-print for instruction developed in the preceding stages, the designer then begins to *develop the instructional units* (lessons). As each lesson is produced, it is tried out on three to five sample students, who are then given that portion of the criterion test that applies. The testing is needed not to assess student achievement but to evaluate the effectiveness of the instruction, *to validate the learning units.* If deficiencies show up, as indicated by poor performance on related groups of items, the lessons are redesigned by resequencing and adopting new learning strategies. The lesson is then revised and retested, using a sample of about ten students. Once again, based on the results with the ten students, the designer will probably have to revise the lesson. If he is reasonably confident of the material at this point, the designer should then give the lesson to at least 30 students, the goal being to have 85 percent of the sample reach the criterion set out in each objective for the lesson. If 85 percent of the 30 students do meet the standards specified by the objectives, the designer can be reasonably sure that 85 percent of future students will also be able to reach the criterion (unless there is a significant change in student characteristics). After a final revision based on this last evaluation, the lesson is considered validated. The same development cycle of develop, test, revise, and retest is followed for each of the lessons in the course.

Finally, *the entire system has to be validated;* all the lessons in combination have to be tested against a population of at least 30 students. Again, the goal is for 85 percent of the sample population to attain the objectives. If the goal is not reached, there would have to be some further revision and retesting. However, at this stage in the development process, there usually is a good chance for each lesson to reach 85 percent the first time.

After completing the testing cycle, the designer is reasonably sure of the success of his system. However, there is one further test to be faced. It is then time to *implement and to field test the system* under actual classroom conditions, with the regular instructional staff conducting the instruction. This, of course, calls for the development of lesson plans, demonstration of the system, and instructor training. Once again, the results of end-of-lesson, end-of-unit, and end-of-course criterion tests are used to evaluate the system. There may still be need for some revision based on the results of classroom experience. *Actually, an instructional system is never complete, because it is a dynamic process, not a static product.* Objectives may change as job content changes. Instructional sequence and strategy may have to be changed because the student population may change. Actually, an instructional system by its very nature involves *continuing and concurrent reassessment of the efficiency and effectiveness of the instruction* based on the analysis of criterion test results.

The foregoing brief overview of the instructional systems development process is pictorially summarized in a flow chart on the following page.

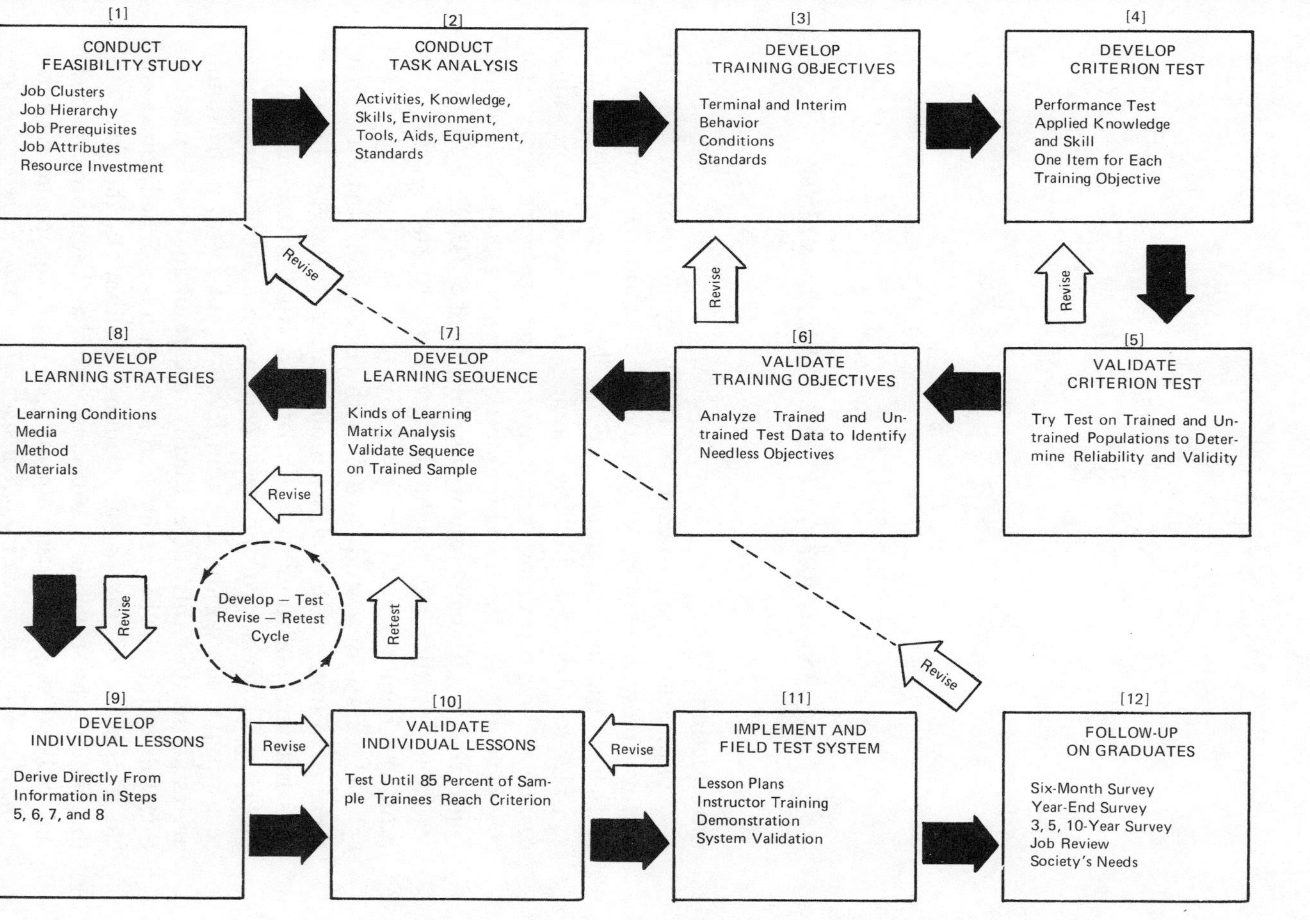

Flow Chart of Training System Development Process

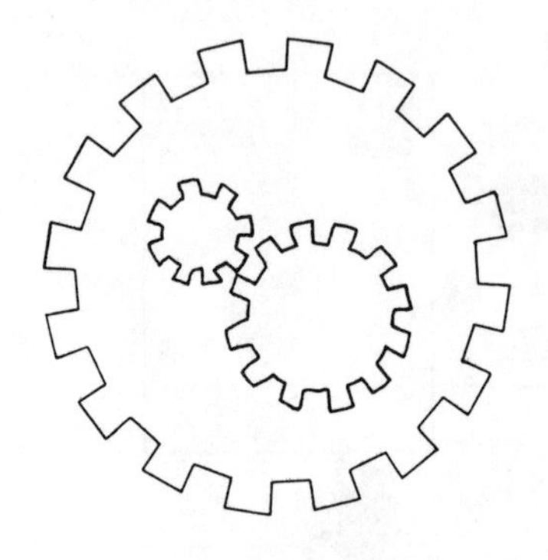

CHAPTER VI

TRAINING PROGRAM FEASIBILITY STUDY

Introduction

Ordinarily, the concept of feasibility is confined to the narrow question of the capability of a school to carry out a particular training program. The feasibility study should also consider the desirability of the training: If a school is to conduct certain technical training, *will it lead to the right kind of a job for the student?* To answer this question many factors have to be considered, and many questions have to be asked. Systematic feasibility studies are absolutely necessary to the success of a training program.

The feasibility study requires three levels of analysis. First, there is an initial selection of an occupational cluster or family of jobs that are horizontally interrelated. Second, a task-related job hierarchy is developed and described in some detail. The vertical relationships among the different levels of jobs within the family are established at this stage. Finally, a careful analysis is made of each job within the hierarchy, and a decision is reached concerning the feasibility of the overall training program. In the following discussion, the feasibility study process has been broken down into several stages for the purposes of explanation and organization. However, in actuality, *all phases of the study are carried out nearly simultaneously because the information developed and the decisions reached at each stage are so highly interdependent.*

Selection of Job Family and Clusters

As stated earlier, vocational course objectives must be described in specific operational terms and must reflect the behaviors desired of the students upon completion of the course. These objectives are derived from statements of the tasks actually required in jobs that are representative of the job families selected for inclusion in the curriculum. The steps in the process of arriving at course objectives are listed below:

1. Select job family.
2. Select set of jobs most representative of each level in the family (horizontal relationships).
3. Develop task-related job hierarchy within the family (vertical relationships).
4. Describe duties involved in each job.
5. Identify tasks required in performance of each job.
6. Identify skills and knowledge required for performance of each task.
7. Derive performance objectives and sub-objectives for individual learning units and lessons.

The present discussion is concerned only with the first five steps in the process. Succeeding chapters will cover steps 6 and 7 in detail.

There are four major sources that can be used to identify and to select the job family for consideration.

1. *Occupational Outlook Handbook*
2. *Dictionary of Occupational Titles*
3. Army, Navy, and Air Force Specialties Training Standards
4. Job Corps Training Standards

There are other sources, of course, but these probably

represent the most highly organized and detailed information available today concerning job families and sub-families, and the specific jobs within each. These sources describe a tremendous range of jobs; however, the purpose of the feasibility study is to select the job cluster that is both practical and desirable. A smaller set of jobs which is representative of the job family must also be selected. That is, *the jobs selected should differ from one another in the kind and level of performance they require and, taken as a group, should include substantially all of the skills and knowledge that are demanded by the job family and appropriate for the training program.* With these end results in mind, the following criteria and general considerations should be used for the initial selection.

Jobs are selected that:

1. In comparison with related jobs, require performance of a wide variety of tasks and a broad range of skill levels. Thus, for instance, the job of auto-body repairman would be chosen on this basis rather than the job of spray-painter, since the auto-body repairman should be able to perform all the tasks of a spray-painter and other more advanced tasks as well.

2. Require an appropriate amount of vocational training time. That is, jobs requiring only a rather short period of training for most students probably should not be listed. Similarly, jobs which require training time beyond that expected to be available should not be included.

3. Have entrance, apprenticeship, or on-the-job training requirements which can be met better as a result of vocational training. Thus, jobs for which the trained graduate could substantially meet the entrance requirements, or would be allowed to progress more rapidly through apprenticeship and additional training programs, would be favored for selection. Jobs which could be entered only after long service in another job, or only after an extended fixed period of apprenticeship or additional training, or only by meeting requirements beyond the control of the training agency

would be less desirable candidates for selection. This does not imply that the content of vocational training should include only that which pays off immediately in a job. Rather, the intent is to foster meaningful and lasting vocational rewards for the student who performs successfully in training.

4. Are appropriate with respect to the cost, size, support requirements, and expected usage of training facilities and training equipment. This consideration may limit offerings in the computer servicing field, for example, to tasks not requiring frequent access to a large digital computer.

5. Are predictable with respect to the skills and knowledge which will be required in the next five to ten years. Of course, changes may take place unexpectedly in any vocational area. If, however, an occupation is undergoing changes due, for example, to mechanization, introduction of new procedures, or revision of job structures, then effective training for that occupation may change radically. In such a case, training plans can be prepared when the performances, skills, and knowledges can be identified, but not before.

6. Have favorable employment expectations (particularly for the local region in question) in the time period for which training is being prepared. The major source for data needed to evaluate future employment opportunities is the *Occupational Outlook Handbook.*

7. Are *not*, in themselves, "dead-ends." Every job, starting with the lowest level, should be a prerequisite for and lead directly to a series of higher-level jobs. Each succeeding level should be an identifiable job, in demand on the labor market. Although each job in such a "career ladder" requires an increasing degree of skill and knowledge, each should be an "entry-level" job in relation to the next higher rung on the ladder. Ideally, *the "career ladder" should begin at the semi-skilled level and proceed upwards, job-by-job, all the way into the ranks of the professions.*

The field of electronics technology will serve to illustrate the outcomes of this initial selection process and the subsequent identification of a job family with its structure of sub-families. While the data shown are factual, the material presented should be considered as *illustrative* only, and should not be taken as definitive.

Analysis of the task content of the many and varied jobs in the field of electronics technology will result in the identification of a group of interrelated major occupations. The jobs should then be classified, and grouped according to the similarities of the skills and knowledge required. Next, an analysis should be carried out to *relate the job groups both horizontally and vertically,* based on the levels of skills and knowledge required. "Clusters" or sub-families of interdependent jobs will then begin to emerge. *The result is a "network" of interrelated jobs that comprise the job family.* The structure of the job family can then be charted as shown in the accompanying illustration.

Development of Task-Related Job Hierarchy

The next step involves the selection of a smaller set of jobs which is most representative of not only the emerging clusters or sub-families, but of the job family as a whole. Care must be taken to select jobs that differ from one another in the level of performance required and, when taken as a group, include substantially all of the skills and knowledge that are both appropriate for the training programs and demanded by the job family.

The job hierarchy chart on page 60 illustrates *a sequence of training courses based on representative jobs that are, in fact, career milestones and exit points within the job family.* The content of each succeeding training course is derived directly from the skills and knowledge required by each of the representative jobs that make up the career ladder. The examples given in the chart include only jobs at the technician level that are appropriate for either secondary school courses or for non-degree post-secondary programs.

To carry the idea to its logical conclusion, however, would require the hierarchy to be extended upward to include non-

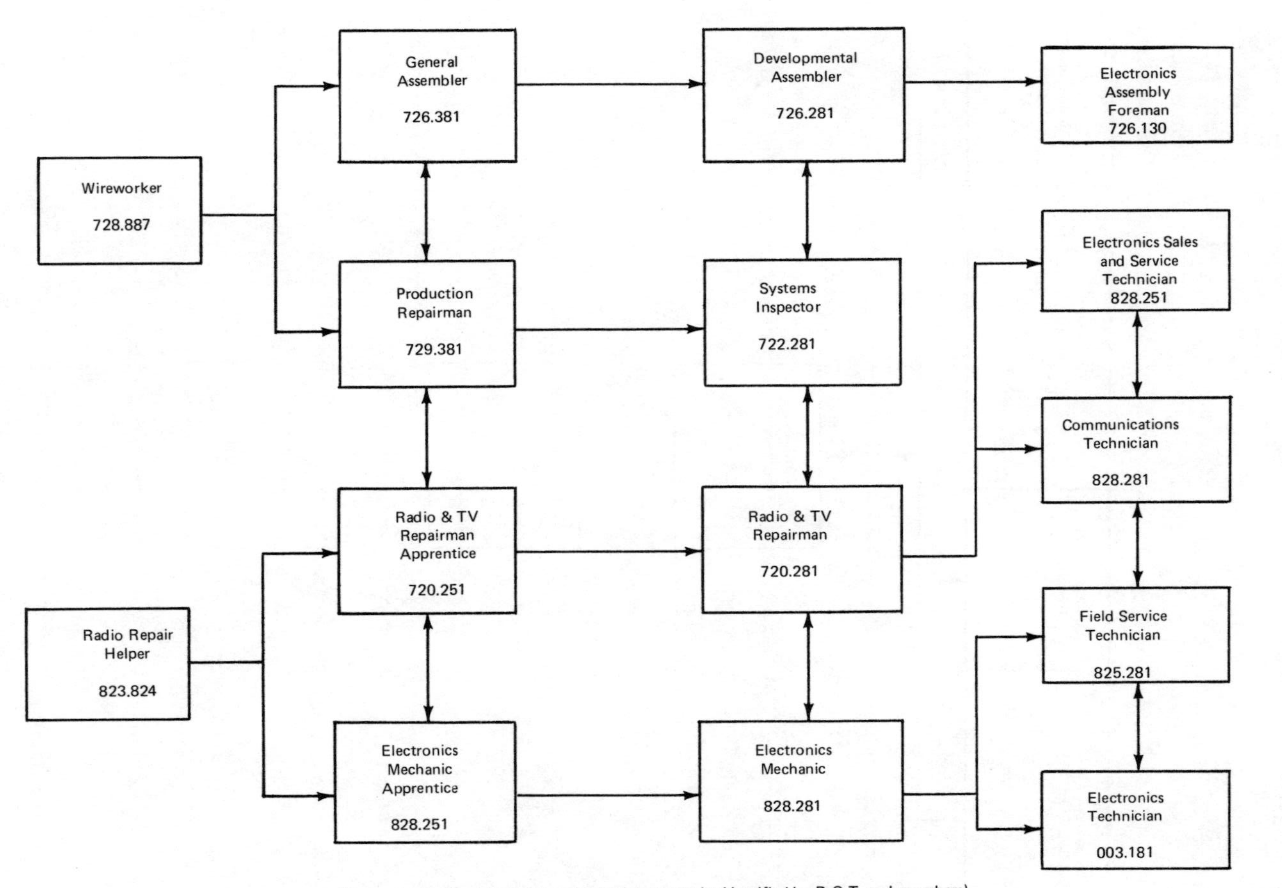

Electronics Family Job Structure (major job categories identified by D.O.T. code numbers)

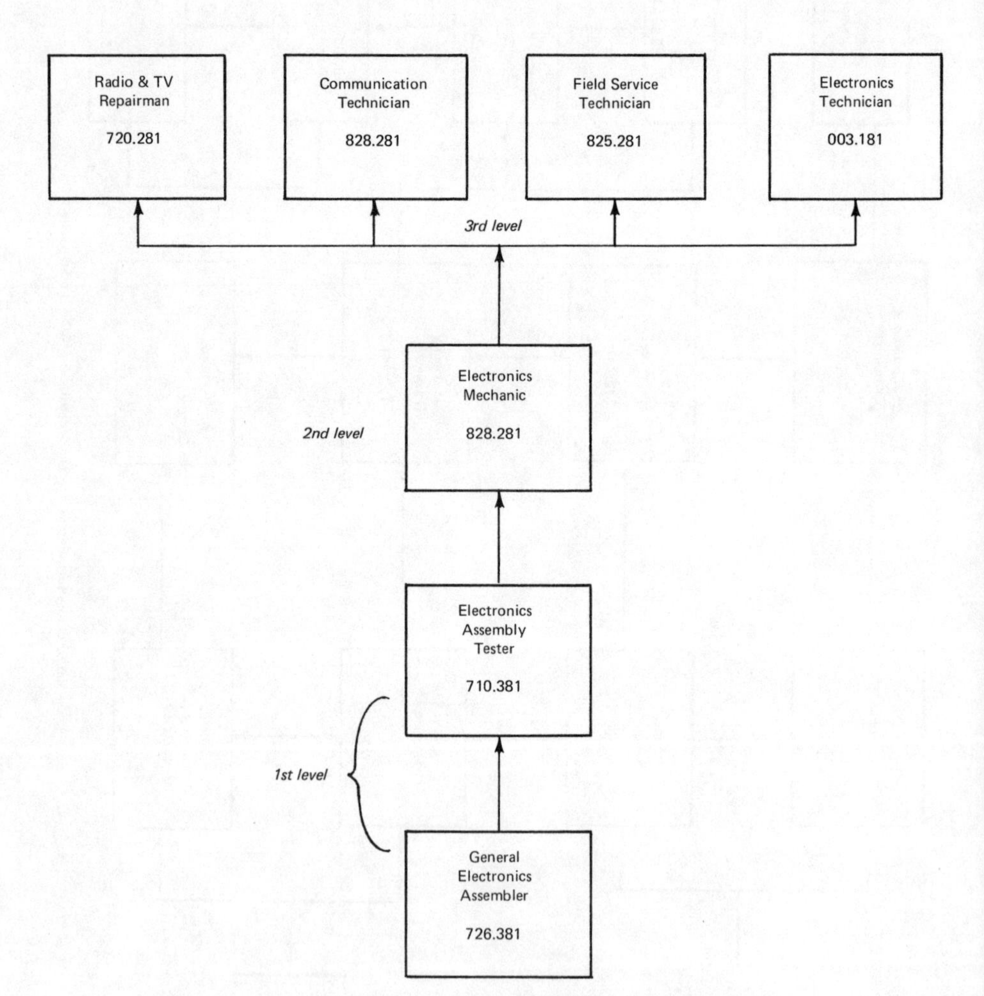

Job Hierarchy for an Electronics Technician Curriculum
(most representative job categories)

degree engineering assistants, two-year associate engineer degrees and, ultimately, four-year electronics engineer degrees. Thus, *completing any one course serves both as an end in itself and as a means to a more distant goal.* Moreover, organizing a curriculum in this manner allows a student, when conditions compel him, to leave the program at an exit point that provides him with a marketable job capability. Likewise, the student, when conditions permit, could later re-enter the program at the level at which he left it.

Each succeeding course qualifies the student as an entry-level worker in that particular job and, at the same time, qualifies him for the follow-on higher-level course. Under these circumstances, a student can proceed through increasingly demanding occupational training until, either by inclination or aptitude, he decides he has gone as far as he can. *Each student, in effect, finds his own level of competence and, regardless of the level attained, he will have at least a marketable job capability.* In contrast, most occupational education (and education in general) today is organized on an all-or-nothing basis with little or no reward for partial completion of a curriculum. The reward of meaningful employment or further education is withheld unless the entire long-range program is completed and its accompanying degree is awarded.

Because each course is based on the single most representative job for a particular level of highly similar skills and knowledge, completion of that course qualifies the graduate for a whole cluster of closely allied jobs. The task-related job cluster list below gives examples of sub-families formed of closely related occupations. The occupational titles and code numbers are those found in the *Dictionary of Occupational Titles.* The representative job which determines the title and the content of the training course requires very similar capabilities and, in many cases, the very same capabilities at a higher or equal level required by the other jobs listed for the cluster. Thus, *qualifying for the title job of the cluster qualifies the student for all the other jobs in the cluster.*

Task-Related Job Cluster List

1. *General Electronics Assembler (726.381)*

    Coil Builder (724.781)
    Bench Assembler (726.781)
    Electrical Assembler (827.884)
    Electronics Assembler (726.781)
    Chassis Assembler (726.884)
    Wireman (726.884)
    Electro-Mechanical Assembler (726.781)
    Wireworker (728.887)
    Telephone and Telegraph Equipment Assembler (722.381)

2. *Electronics Assembly Tester (710.381)*

    Assembly Adjuster (720.884)
    Aligner (720.884)
    Chassis Inspector (720.687)
    TV Tube Tester (720.684)
    Tester, Electrical Continuity (729.684)
    Tester, Components (726.687)
    Electronics Assembler and Tester (710.381)
    Parts Changer (729.381)
    Rework Assembler (726.281)
    Production Repairman (729.381)
    Television Chassis Inspector (720.687)
    Television Chassis Aligner (720.884)
    Components Inspector (726.687)

3. *Electronics Mechanic (828.281)*

    Systems Inspector (729.381)
    Electronics Assembler, Developmental (726.281)
    Radio Mechanic Helper (823.884)
    Electronics Maintenance Man Helper (828.281)
    Computer Mechanic Helper (828.281)
    Radar Mechanic Helper (828.281)
    Communications Mechanic (828.281)
    Components Inspection Technician (828.281)
    Television Mechanic Helper (823.884)
    Television Installer (823.781)

4. *Radio and Television Service and Repairman (720.281)*

Radio Repairman (720.281)
Radio Mechanic (720.281)
Television Repairman (720.281)
Electronics Technician Apprentice (828.281)
Public Address System Serviceman (720.281)
Automobile Radio Man (720.281)
Tape Recorder Repairman (720.281)
Audio-Visual Equipment Repairman (729.281)
Audio-Visual Repairman (729.281)

Furthermore, the horizontal relatedness among the jobs within a cluster proceeds vertically between clusters as well. The jobs in the lower-level clusters encompass skills and knowledge that are basic to the tasks required in the higher-level clusters. Also, many of the lower-level tasks continue into the higher-level jobs, but with more stringent standards and as part of broader areas of responsibility. *The ancillary jobs, as well as the primary jobs, can thus provide the same opportunity for moving up the career ladder.*

The same sort of career ladder can be structured for every occupational area. For example, the social service career curriculum might be organized as follows: Social Service Aide → Case Worker Aide → Community Health Worker → Community Health Technician → Social Worker Assistant → Social Worker Technician → Social Worker. The food services curriculum might be organized as follows: Kitchen Aide → Cook's Helper → Assistant Cook → Cook → Head Chef → Dietary Assistant → Dietitian. Dentistry education could be structured as follows: Dental Aide → Dental Assistant → Dental Hygienist → Dentist. Studies leading to professional positions in business management might be organized around the following hierarchy of jobs: Business Office Aide → Posting Machine Operator → Bookkeeper → Accountant → Comptroller. The above career ladders are mostly conjectural and are meant only to be *suggestive* of the kinds of structure that should be considered. A great deal of carefully detailed analysis has yet to be carried out before education for the majority of the career fields can be recognized in this manner.

Organizing curriculum content in such a hierarchy of task-related job clusters goes a long way towards minimizing the specificity vs. flexibility dilemma of vocational education. The skills and knowledge imparted by the training are job-specific, but at the same time, they are highly generalizable, both horizontally and vertically. As the student progresses upward in the hierarchy, the scope of the job increases; and, thus, he is provided with an increasingly broader base of generalizable capabilities. *The potential for the cross-transfer of skills and knowledge increases in direct proportion to the level of the job in the hierarchy.* Starting at the lowest entry-level job, the number of generalizable capabilities increases with each higher-level job, thus building an inverted pyramid of transferable skills and knowledge.

Preparing Job Descriptions

During even the initial stage of the feasibility study, an elementary description of the jobs under consideration is needed to organize the relevant information. The purpose of a job description is to provide data which are useful in defining the performance required of an incumbent. The analysis of the content of the job provides the basis for the identification of the requisite capabilities, and thus helps the curriculum developer to assess the characteristics of the job and its relatedness to other jobs.

Only rather simple and general descriptions are needed at first but, *as the study progresses, increasingly detailed and exact job descriptions have to be developed to support the increasingly rigorous assessment of the feasibility.* The job descriptions are continuously expanded and refined until, at this point in the study, they are essentially complete, requiring only a final review and edit. Sample job descriptions for two levels of training in the field of electronics technology can be seen in Appendix A.

The initial section of the job description, which is devoted to *Defining the Population,* distinguishes between the jobs having similarities to be included and excluded. A brief general description is given of the general characteristics of job incumbents along with general information about the industry which helps to delineate the tasks. A *Statement of Mission* identifies the general objectives of the job and may also define alternative objectives, operational modes, and hierarchies of goals. It sets general criteria by which one can judge performance and sets the objectives toward which all tasks are aimed. A *Duties* section identifies sub-operations of the mission and serves as an important organizational aid for defining the tasks. Duties also indicate sequences, time phases, and categories of operations. They are the major steps in the regular sequence of job performance. The section labeled *Contingencies* identifies conditions under which the job is to be performed—the usual and the unusual.

A *Basic Tasks* section provides a list of specific statements of performance that make up the job content. *The listing of the basic tasks is critical to the process of identifying the essential skills and knowledge and to the assessment of program feasibility.* Moreover, the identification of the basic tasks lays the groundwork for the detailed task analysis that is the next step in the systems development process.

When the job hierarchy and accompanying job descriptions have been completed, the next step is to carry out the final and more detailed analysis of the factors affecting each element of the proposed training program. In some ways this final review is merely an affirmation of the continuous assessment process that was tentative at the outset but which became "increasingly rigorous" as the analysis progressed. However, *it is important at this stage to conduct a complete and more detailed review of the earlier findings and feasibility decisions.* The following sections cover the steps in that review process and the subsequent final decision.

Analysis of Job Prerequisites

A key question is: *What are the special prerequisites of this job cluster?* This question breaks down into eight important sub-areas, each of which adds to the final decision. The job prerequisites to be considered include the following factors:

1. Vocational skills
2. Educational level
3. Licensing, certification, and entry testing
4. Union membership
5. Previous job experience
6. Age
7. Physical ability
8. Sex

Can enough training be provided to develop the skills demanded by the prospective employer? It is important that the skill levels required for job success are clearly identified.

The feasibility study should also uncover the current educational demands of the industry. Some industries, in cooperation with unions, have established graduation from high school as an educational requirement. Some even require post-high school educational attainment. If an industry has taken 15 or 20 years to raise its educational demands, it will not lower the bars overnight.

A closely allied area is the licensing, certification, and entry testing requirements. It is important to know what these requirements are and what the possibility is for students to meet and pass these requirements.

The next area of interest is union membership requirements. Does union policy restrict recruiting and membership in the particular trade under consideration? Does this union bar minority races from union membership? A close investigation of union problems in a particular career field should be made before offering a training program. Some unions are very eager to broaden their membership. They may be able to provide assistance in many areas of your training and placement efforts. Several excellent training programs have failed in the past because union practices were not taken into account.

The age, physical, and health requirements of a given job must always be considered. Jobs that would be open to 21- and 22-year-olds are closed to younger students. It may be important to determine the age and physical demands of a job before training is started. Many jobs are very demanding physically. Some thought might be given to developing training programs that can be used by both men and women, but is there any real opportunity for women in the job cluster?

Analysis of Job Attributes

The second most important question is: "*What does this job have to offer the graduate?*" Again, this question can be broken down into further considerations:

1. Entry salary or wage scale
2. Salary after five years
3. Advancement opportunities
4. Personal considerations
5. Labor market considerations, including significant regional placement opportunities (How many jobs available where?)
6. Technological considerations, including where the field will be five and ten years from now
7. Training opportunities
8. Job image

Not only the starting wage scale should be considered but also the expectations after five years' experience. Many jobs have minimum starting wages but pay good wages after a few years' experience and further training. Other jobs pay what appears to be acceptable wages from the very start. However, there are dead-end jobs that do not pay more with increased experience. Do not accept or reject a given job classification based on starting wages. Consider both the beginning wages and potential wage increases. There is more to job advancement than just wages. Some youths may be content with a well paying job that offers little job progression. Other men and women want the opportunity to advance in skills and responsibilities.

What are the personal considerations concerning the jobs? Is extensive travel required? Is the job blue-collar or white-collar? Is it an outside or inside job? Are there any odd working hours or shift requirements? A close investigation of the actual working conditions will reveal the personal aspects of the job that must be considered in the feasibility study.

Even a cursory look at the labor market will reveal the important facts that need considering before deciding to provide training for a given job cluster. The main consideration is to determine how many jobs are available where. The electronics industries are concentrated mostly on the West Coast and in the Northeast. Automobile mechanics are in demand nationwide. Some jobs have only seasonal requirements, others have regional demands. Others are in demand only in urban or rural areas. The size, scope, and location of the labor market must be determined.

Where will this career field be five or ten years from now? Is the job on the way out, or is it a part of a new and emerging industry? Is it part of the paramedical field that is expanding because of recent government legislation? Is it part of the communications, plastics, or electronics industries? Training should be for the future, not just for the present. The training program should help not only the students but also industry as a whole by training for jobs in new or expanding fields. Training programs should not be started for dying industries.

Another question that needs investigation in this area is that of job advancement training opportunities. How much training should be provided to insure that the student can take advantage of further training? How much "education" needs to be added to the training provided? Should the goal be only the first, entry-level job, or for the higher-level, higher skilled job?

The final question to ask is: "What do students think of this job?" Is there sufficient student demand for the training? This question may be hard to answer in absolute terms, but it must be considered. There are many training programs that have high dropout rates. One reason for this has been the lack of appeal of the job. It would be wise to consider this aspect before making a final decision to develop certain training areas. The more costly the training, the more important this aspect becomes.

In the past, some career fields have been oversold, only to have a high dropout rate and low actual job placements. It is possible to determine student likes and dislikes. Surveys will help determine whether to start on a new training program. The survey should sample an adequate number of students, and it should be as unbiased as possible. It should not attempt to sell a new or existing program.

Analysis of Training Investment

A third question that must be answered is: "*What is it going to take in terms of time, manpower, facilities, and materials to develop and conduct this training?*"

1. Statistical information
   a. Student flow
   b. Entry rate
   c. Training requirements
   d. Length of course
2. Equipment requirements
3. Facilities requirements
4. Instructor requirements
   a. Instructors available
   b. Instructors required
   c. Instructor training required
5. Cost analysis
   a. Start-up costs
   b. Cost of changing a present course
   c. Off-site cost comparison

Statistical, financial, and manpower information is needed to answer this third question. Without this information, it is possible to set up programs that are much too expensive to maintain. Appliance repair, television repair, business machine repair, and reproduction and printing training are relatively costly programs to develop and to operate. Not all training programs are expensive, but limited budgets may demand cost consciousness.

Who else is training toward the same job vacancies? Without at least a quick survey of the Departments of Labor and of Health, Education, and Welfare plus state and local agencies, a school may end up training men for jobs that two or three other local agencies are already training toward. There is nothing wrong with this duplication unless the total effort leads to a surplus of trained graduates trying for the same job vacancies.

This is really a two-part question. If other agencies are already training toward the same job vacancies, perhaps advantage should be taken of these other training opportunities. This can be done in several ways. One way is to send the students to these other training facilities. Another way is to develop cooperative work-study programs to take advantage of industry on-the-job training. Still another way is to use only parts of an existing program and modify or expand it to meet the specific local needs. Some schools are doing this to a degree already, but not to the full extent possible. In fact, there has been a tendency to reinvent the wheel by redesigning existing programs that are already effective. In summary, make sure that training does not lead to a surplus of trained men or women, and capitalize, to the maximum extent possible, on nearby existing training facilities and training development activities.

Feasibility Decision

It is important to ask the right questions, but it is just as important to ask them in the right order. Covering the more important considerations early in the study will often save time and money. The following list gives the recommended sequence. *An unfavorable response to any of the first six questions should probably lead to an unfavorable decision.* Once past the first six questions, you have to weigh the pros and cons of each of the remaining questions as listed below:

1. LABOR MARKET
2. STUDENT DEMAND
3. AGE
4. RACE
5. SEX
6. UNION MEMBERSHIP
7. Entry-level salary or wages
8. Salary after five years

9. Advancement opportunities
10. Physical requirements
11. Licensing, certification, and entry test requirements
12. Educational attainment
13. Vocational skills required
14. Previous job experiences required
15. Training costs
16. Technical considerations
17. Training opportunities
18. Personal considerations

There is no simple or exact formula for arriving at the correct decision. However, this chapter has tried to call attention to some of the more important factors to be considered before making a final decision to proceed with developing and conducting a given training program.

Periodic Program Review

Every training program should be reviewed every two or three years because of the dynamic nature of American society today. A program that was not feasible because of racial considerations last year may be feasible this year. On the other hand, a program based on a sound decision made a few years ago may no longer be necessary nor even fit the current needs of the American industrial scene.

First of all, "*Is the training effective?*" This can be answered only by a thorough analysis of the current training. The following factors must be considered:

1. Statistical data
   a. How many students have been trained
   b. Cost per student
   c. Length of course
   d. Entry rate
2. Curriculum materials
3. Training methods
4. Training staff

5. Testing results
6. Placement success
7. Job retention rate
8. Employers' opinions
9. Graduates' opinions

Be sure to take a hard look at graduate placement data, for they are important in evaluating program success (see Appendix O for follow-up study form). Poor placement figures may be a strong indication of a poor training program. However, do the figures square with employers' opinions of the capability of the graduates? Are the placement data at odds with all other indicators of program success? If graduates are having trouble finding jobs, perhaps the job market has changed, or the fault may very well stem from a poor or non-existent school placement and referral service. Numerous studies have shown that *an active placement service must be maintained by the school itself, if the success of even a superior training program is to be measured in terms of the number of graduates placed in jobs.*

A third question is: *Should the present training be modified to make it more effective?* This leads to two more questions: What will it cost to change methods? What are short-term and long-term savings, if any, of a redesigned course? Another question could be: *Should the present training program perhaps be eliminated?* And several others are:

What will it cost to continue?

What will it cost to stop training?

All these questions should be answered as part of a review of existing training programs.

Make sure a training program is accepted, maintained, altered, or rejected for the right reasons. Good feasibility decisions are an absolute necessity to the success of occupational education. *The effects of a poor feasibility decision may take years to overcome.*

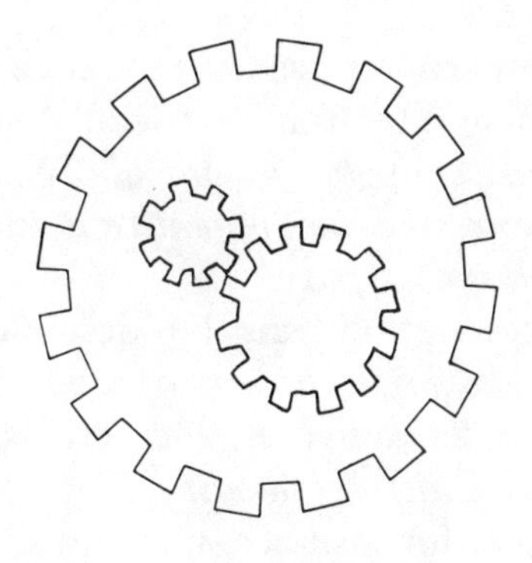

CHAPTER VII

TASK ANALYSIS

Introduction

Once the feasibility study has been completed and the decision has been made to develop a certain training curriculum, the next step in the process is system design. *During the design phase, the prototype model* of the training system will be produced, and *all the major decisions concerning training objectives, content, sequence, method, media, and evaluation* will have been made. In addition, the validation device containing the performance criteria against which the system will be evaluated will have been developed.

Job/Task Description

Task description is the foundation upon which the entire system is built. A thorough and accurate job/task description is absolutely essential to the entire structure. Because it provides the substance for the content of training, task description suggests the sequencing and form of training, and also serves as a statement of the performance criterion which will be used in evaluating both the training and the students. *Task description is virtually the fundamental source of training objectives.* The degree of detail needed for a job description in systems development is much greater than that normally associated with job classification manuals and training literature. Generally, *a task description should be detailed enough to provide minimal step-by-step directions and guidance that an apprentice in training could follow to complete the task successfully.*

A job/task description can be described as *a summary statement of the behavioral content of a job* broken down into component duties, tasks, activities, and actions. *A duty is a major sub-division that has a distinct identity within the overall job.* The job of auto mechanic, for instance, includes such duties as engine tune-up, engine overhaul, and wheel alignment.

Duties are in turn composed of several distinct tasks; for example, cleaning the spark plugs, replacing the points, setting the timing, and adjusting the carburetor are tasks under the general duty of engine tune-up. *A task is a logical and necessary step in the performance of a duty—usually a fairly long and complex procedure.*

Tasks are in turn made up of a series of activities with a common purpose that occur in close sequence. Removing the old points, installing the new points, and adjusting the dwell are activities under the task of replacing the points. *An activity is a logical and necessary step in the performance of a task—usually a relatively short and simple procedure.*

Activities are in turn made up of a series of actions or manipulations that fall closely together and have a common purpose. For instance, loosening the lock nut with a wrench, attaching a dwell meter, reading the dial on the dwell meter, and setting the point gap are some of the actions under the activity of adjusting the dwell. *Actions or manipulations are short, simple operations that are frequently common to many different activities and involve using tools, devices, controls, and simple test equipment.*

A task description is usually developed in three stages. The duties that fall under the job are first outlined. Next, the tasks that fall under each of the duties are listed. In most cases, the duties and tasks can be determined fairly accurately by consulting training materials, manuals, and the literature. However, task description *at the activities level can only be developed by actual on-the-job observation and interview.*

The activity content of each task should be described in the following terms:

- The cues, signals, and indications that call for the action or reaction
- The control, object, or tool to be used or manipulated
- The action or manipulation to be made
- The cues, signals, and indications (feedback) that the action taken is, or is not, correct and adequate

The task description must also identify *working conditions* and *environmental situations* which might affect job performance adversely. Any special precautions and possible alternatives must be listed for each task. Also, *special tools,* test equipment, and manuals should be identified. And lastly, the task analysis must identify *standards of performance.* If there are critical time and accuracy demands (and there usually are), they must be listed. Performance standards such as these provide the criteria for proficiency measurement.

There are many forms that will serve to organize the collection of the task information, but the samples shown on the next two pages are recommended for general use. Although the information on the forms is realistic, it is not factual, and is for illustrative purposes only. Note that all the information needed for a thorough task description can be summarized on this form. Moreover, the format organizes the information in a logical manner that prepares the way for the follow-on task analysis.

There are many possible sources of job description information. Some of the more common are training literature, manuals, course outlines, textbooks, and occupational qualifying exams—all of which may serve as a good starting point. After compiling a consensus of the training literature, the next step should be to consult subject matter experts from training schools and from the occupational field. A review by a subject matter expert can be very helpful. However, *the best and final authority is a sampling of*

Page *1* of *25* Pages

JOB: Statistical clerk (Sample Task Description Form)

DUTY: 5 Perform simple arithmetic operations on calculator

TASK: 5.1 Add column of positive numbers with decimals

ANALYST: R. Ewing DATE: 6/1/71

INCUMBENT S.Church COPY: 1

ACTIVITY Code + Activity Statement	Time Scale (minutes)	ACTION STATEMENT Code + Action + *Item Acted Upon* + Modifiers	ACTION DETERMINANT(S) Plus Info Needed to Act	INDICATION OF RESPONSE ADEQUACY Plus Info Needed to Determine Adequacy	REMARKS Reference to Alternate or Emergency procedures (AP, EP); Comments and Definitions; Precautions
5.1.1 Set up machine for addition of numbers with decimals	00	. . . .	Desire or instruction to perform the task	As indicated below	Assume a prior orientation to machine
		5.1.1.1 *Depress* DIALS CLEAR and KB CLEAR	SP (Standard Procedure) to clear keyboard and dials	Zeros appear in all dials. No key is depressed except blank or zero keys	If numbers still appear in MULTIPLIER dials; clear by pressing CLEAR MULT and repeating Element I If numbers still appear in UPPER or LOWER dials; raise LOCKS for all dials and raise CONSTANT MULTI-PLIER lever
	01	5.1.1.2 *Depress* and hold upper CARR SHIFT	SP	Carriage in No. 1 position	

(Sample Task Description Form)

JOB: Welder

Page *4* of *30* Pages

DUTY: 8 Operate mobile motor generator

TASK: 8.3 Start motor-generator

ANALYST: R. Ewing DATE: 6/1/71

INCUMBENT: J. Brown COPY: 1

ACTIVITY Code + Activity Statement	Time Scale (minutes)	ACTION STATEMENT Code + Action + *Item Acted Upon* + Modifiers	ACTION DETERMINANT(S) Plus Info Needed to Act	INDICATION OF RESPONSE ADEQUACY Plus Info Needed to Determine Adequacy	REMARKS Reference to Alternate or Emergency procedures (AP, EP); Comments and Definitions; Precautions
8.3.1 Adjust control panel	00	. . . .	Desire or instruction to perform the task	As indicated below	Assume familiarity with control panel
		8.3.1.1 *Press* POWER ON button	Standard Procedure (SP)	Motor starts and makes audible hum, pilot light comes on	Avoid starting with bus bar covers off: personal hazard
		8.3.1.2 *Turn* AC voltage control knob as far as required	SP	AC voltmeter aligns to 117 ± 4 volts	
	02	8.3.1.3 *Turn* FREQ adjustment screw (screwdriver)	SP	FREQ meter indicates 60 ± 5 cycles and holds steady	If frequency fluctuates, more than ± 5 cycles, turn FREQ adjust back to zero; then slowly readjust to 60 cycles If still fluctuating more than ± 10 cycles, shut down generator at once to prevent damage
	04	8.3.1.4 *Depress and hold* POWER OFF button as long as required	Emergency Shut Down for fluctuating frequency	Motor stops, no audible hum, pilot light goes out, meter zero	

the on-the-job worker. Foremen, journeymen, and apprentices should be observed and interviewed on the job. Remember, however, that most training programs should be aimed toward entry-level jobs. Therefore, most emphasis should be placed on *observation and interview of the apprentice or entry-level worker to find out what he actually does on the job*—what is actually demanded by the job, not what someone thinks the entry-level worker ought to be able to do. Many training programs for entry-level workers include a great deal of unnecessary material that concerns tasks performed only by the highly skilled journeymen. The apprentice usually forgets such "nice-to-know" information rather rapidly and soon loses his unused advanced skills. Even if he were to retain the skills and knowledges until he actually needed them, he probably would find them outdated by then. Moreover, many of the more advanced skills and knowledges are often best acquired on the job or only after considerable experience at the apprentice level.

Job/Task Analysis

A task description is a list of job activities couched in essentially physical terms. In contrast, *a task analysis lists the behavioral characteristics of the job requirements.* Systematic study of the behavioral requirements of the tasks is needed *to determine the knowledge and performance content of the job.* Even though the two activities, description and analysis, are here treated separately, both kinds of information, the physically descriptive and the behaviorally analytic, are closely related with no clear dividing line between them. Usually, however, the detailed behavioral analysis is done only after the task description phase is complete.

The first step in task analysis is to identify the kinds of performance capabilities demanded by the tasks. *Each task must be analyzed to determine what kinds of learned performance are involved, for this information is the basis for all instructional system design decisions.* The selection of appropriate objectives, content, sequence, method, media, and criteria depends on the correct identification of the capabilities needed to perform the tasks. Most of the action verbs used in task description tend to fall within one of the specific types of learning outlined in Chapter III. Of course, it is not always easy to categorize the kind of activity, but careful analysis will usually identify the performance needs of the task in question. Table 3, which follows, lists some common action verbs that can generally be associated with specific learned capabilities.

The list is not all-inclusive, nor are the words always mutually exclusive; however, reference to the list should help in classifying performance. Actually, an agreed-upon list of action verbs, each assigned to a specific category of performance, should be drawn up ahead of time. Using only agreed-upon verbs during the task description phase will make the later analysis much easier.

Table 3

Action Verbs Related to Specific Kinds of Learning

Specific Responding (producing a single, isolated response)		**Motor Chaining (producing a sequence of motions)**	
to associate	to recognize	to activate	to stencil
to give a word for	to repeat	to adjust	to trace
to grasp (with hand)	to reply	to align	to tune
to hold	to respond	to close	to turn off-on
to identify	to rotate	to copy	
to indicate	to say	to (dis)assemble	
to label	to set	to (dis)connect	
to lift	to signal	to draw	
to locate	to slide	to duplicate	
to loosen	to tighten	to insert	
to move	to touch	to load	
to name	to turn	to manipulate	
to pick up	to twist	to measure	
to place		to open	
to press		to operate	
to pull		to remove	
to push		to replace	

(Continued on Next Page)

Table 3
(Continued)

Verbal Chaining (producing a sequence of words)	**Discriminating (identifying two or more stimuli)**	**Classifying (using concepts)**
to cite	to choose	to allocate
to copy	to compare	to arrange
to enumerate	to contrast	to assign
to letter	to couple	to catalogue
to list	to decide	to categorize
to quote	to detect	to characterize
to recite	to differentiate	to classify
to record	to discern	to collect
to reiterate	to distinguish	to divide
to repeat	to isolate	to file
to reproduce	to judge	to grade
to (re)state	to match	to group
to transcribe	to mate	to index
to type	to pair	to inventory
	to pick	to itemize
	to recognize	to order
	to select	to rank
		to rate
		to reject
		to screen
		to sort
		to specify
		to survey
		to tabulate

(Continued on Next Page)

Table 3
(Continued)

Rule-Using (using principles)		**Problem-Solving (combining two or more principles)**
to anticipate	to expect	to accommodate
to calculate	to explain	to adapt
to calibrate	to extrapolate	to adjust to
to check	to figure	to analyze
to compile	to foresee	to compose
to compute	to generalize	to contrive
to conclude	to illustrate	to correlate
to construct	to infer	to create
to convert	to interpolate	to develop
to coordinate	to interpret	to devise
to correct	to monitor	to diagnose
to deduce	to organize	to discover
to define	to plan	to find a way
to demonstrate	to predict	to invent
to design	to prescribe	to realize
to determine	to program	to reason
to diagram	to project	to resolve
to equate	to schedule	to study
to estimate	to solve	to synthesize
to evaluate	to translate	to think through
to examine	to verify	to trouble shoot

There is no clear division between job knowledge requirements and job performance requirements; and, in fact, both kinds of information may be developed simultaneously. Any listing of knowledge needed on the job reveals the close relationship between knowledge and performance. Each type of learned capability has a supporting knowledge structure that is specific to its needs.

Job knowledge is the minimum information about existing stimulus conditions, the desired goals or end results, and the means and methods for reaching the goals *needed by the worker to insure success in performing a task.* Knowledge requirements fall into three general categories: (1) The knowledge content itself, in the form of a concept, rule, or principle. An example of such knowledge would be "Use care when tightening bolts, don't over-tighten them." (2) Perceiving and identifying when and where the knowledge has to be applied in the work situation. The student should, for example, know that tightening the headbolts on an engine is a very critical operation in which over-torquing or under-torquing can cause serious damage to the engine. (3) The practical application of the knowledge in the situation. The student should know, first of all, that he ought to use a torque-wrench when tightening headbolts; and second, he should know how to use the torque-wrench.

The following is a sample listing of categories of knowledge and information that may prove useful for task analysis:

1. The nomenclature and locations of parts, tools, equipment, etc.
2. General information about common tools and test equipment
3. The clues, signals, and indications to look for
4. The interpretation of symbols, signals, or cues
5. The expected outcomes of actions and decisions
6. The steps in standard and emergency procedures
7. The precautions to take
8. The rules for specific mathematical calculations
9. Methods (rules) for preplanning work
10. Methods (rules) for following strict procedural guides
11. Strategies for problem-solving, diagnosing, and trouble shooting
12. Strategies for improvising and inventing when necessary

The above list may not include all the specific types of knowledge needed, but it does outline the general areas of knowledge that have to be dealt with. Note how the categories of supportive information and knowledge follow closely the hierarchy of learned performance capabilities discussed under the seven types of learning.

There is a tendency on the part of some training specialists to over-emphasize and over-inflate the background knowledge and theory content of training programs. A great deal of "nice-to-know" material included in many courses is not really needed to do the job. In addition, many courses designed to prepare entry-level workers include advanced skills and knowledge that only the journeyman needs. These advanced skills and knowledges, as pointed out earlier in the chapter, are soon lost to the apprentice through disuse. *The supporting knowledge structure should include little information that is not specifically required for task performance.* There are times, of course, when some general information is needed for background purposes. However, introductory and background material should be kept to a minimum; and when it is included, the students should not be required to learn it in detail. *Background material should be treated as such—the students should only become generally familiar with it, not held responsible for it.*

There is another type of unnecessary information that is often found in training courses. If dials, meters, controls, and adjustments are labeled, there is no need to spend time learning their names and locations. Directions, procedures, emergency information, and warnings are often printed on the equipment; *thus, the student should not be required to commit posted information to memory.* Likewise, it is enough to know that he must shut down the engine immediately if the Low Oil Pressure warning light comes on—there is no need for him to know the warning light comes on when the pressure drops below 14.7 pounds per square inch. There are other job aids that lower training requirements. If check lists, tables of data, nomograms, manuals, etc., are normally available and used on the job, the student should learn how to use these aids—but *he should not have to learn how to calculate the data or have to remember the steps in a procedure normally available in some sort of job aid.* There is absolutely no justification, for instance, for having to learn how to calculate square roots the hard way when square root tables are always available. *Close observation of actual on-the-job performance and probing interview of the workers will keep unnecessary and "nice-to-know" material at a minimum.*

The task analysis phase of the systems development process is extremely critical. *The validity of course design decisions is directly related to the care and accuracy with which the task analysis is done.* Training programs must be responsive to the real needs of the job. A thorough-going task analysis is the very necessary first step in that direction.

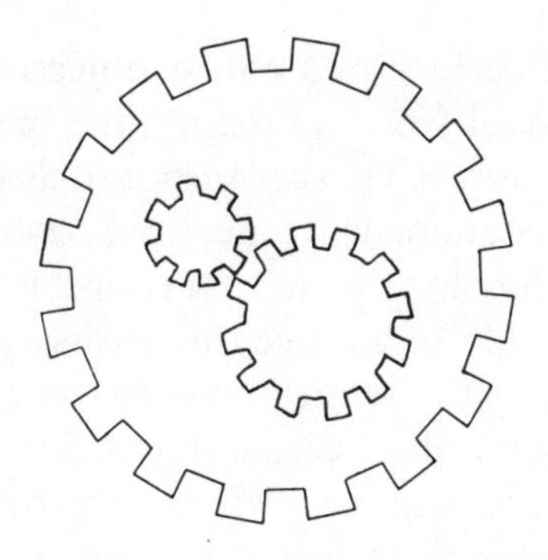

CHAPTER VIII

PREPARING TRAINING OBJECTIVES

Introduction

With the task analysis completed, the next step in the system design process is to develop the training objectives from the information developed by the task analysis. Therefore, the better the analysis, the easier it is to produce good objectives.

There are three main reasons for preparing objectives: *(1) There is a need to know exactly what the student will be expected to do as a direct result of the training. (2) There is a need to know exactly the conditions under which the student will have to perform. (3) There is a need to know exactly what the standards of performance will be.* The instructional systems designer *must have this information to make his decisions* on content, sequence, method, and media. The same information serves as the basis of the criterion tests used to assess the performance of the students and the system itself. Thus, properly stated training objectives provide the very structure upon which the entire instructional system is to be built. Their importance to the course developer is obvious, but there are others who also have a real need for explicitly stated performance objectives.

Training managers also need clear statements of course objectives because they must know exactly what the course is to accomplish. Thus, clearly stated course objectives provide training managers with a basis for evaluating the performance of the system. Managers also have the responsibility for determining requirements for equipment, time, and resources; hence, the clearer the picture they have of what a course is to accomplish, the better they will be able to arrange the necessary support. Training managers also have to relate courses to each other, both horizontally and vertically. Clear statements of objectives for individual courses are also needed to structure an interlocking curriculum. The decisions the managers must make when assigning entering students and when placing graduates are also dependent on the clear statement of performance objectives.

Another who must have access to the training objectives is the instructor. He has to know exactly what the student is to be able to do as a result of his instruction. Moreover, instructors leave and there are always new instructors who need to know what is required of them. Precisely stated performance objectives are the best way to communicate their instructional responsibilities to them. Instructors may change, but the course objectives do not. Unless instructors have access to detailed course objectives, there is an excellent chance that some of the objectives may become overemphasized, underemphasized, or even lost entirely over a period of time.

Finally, and *most importantly, the student himself must have access to the information provided by the training objectives.* Students often spend a great deal of time trying to figure out what the instructor really wants them to learn. The instructor is frequently asked, "Do you really want us to learn this? Will this be on the test? You spent a lot of time on this the last period, and yet it wasn't covered on the test at all." These are simply various ways the students may try to find out what the instructor wants them to learn. Unfortunately, some instructors actually try to prevent students from finding out exactly what is expected of them, claiming their students will therefore learn more! If learning is to be efficient, each student must know the objectives of the training. He should not have to spend time trying to find out what it is he should learn. Without the guidance of clear objectives, he often ends up wasting his time by studying or doing the wrong things. *The strength of the student's motivation and the quality of his performance is directly related to his knowing exactly what he has to do and to knowing how well he has done it when he has finished.*

Specifying Behavior

The task of listing and describing the exact performance capabilities expected of the student is not an easy one. Care must be taken to avoid descriptions that are either too general or too specific to be of any use. The training objectives must communicate clearly to the students, the instructors, and the training managers precisely what is to be learned and how satisfactory achievement is to be demonstrated. Thus, to be meaningful and useful, *training objectives must be described in terms of observable and measurable performance.* To say a student will be expected to "know" something, to "understand" something, or to "appreciate" something at the end of a lesson is an unsatisfactory basis for the planning of that lesson. Such "non-action" verbs furnish merely the starting point for thinking about the instructional problem, but they are entirely inadequate as the basis for making instructional decisions.

"Understanding the theory of electronics" in no way states what the student is to be able to do at the end of the instruction. What does he have to do to show he understands the theory of electronics? Is he going to answer a series of questions on electronics? Is he going to design a radio? Is he going to repair a TV set? *How does the student demonstrate he understands* the theory of electronics? A course in the theory of electronics could last for one month or several years. The course could include a widely different range of content leading to a great variety of knowledge and skills. With a statement as vague and imprecise as "understand the theory of electronics," it is impossible to tell what the objectives of the course really are. *Terms like "know," "understand," "appreciate," etc., are used only in general summarizing statements of the objectives, and always have to be supplemented with specific statements of what the student has to do to demonstrate he understands, knows, or appreciates.* For example:

"Understand the decimal system," breaks down to:

1. Translate any fraction into its decimal equivalent.
2. Translate any decimal into its fractional equivalent.
3. Perform addition, subtraction, multiplication, and division with decimal numbers.
4. Accurately describe and demonstrate each of the above processes to the instructor.

Instead of "know" or "understand," say "write," "recite," "list," "match," "distinguish between," or any of a variety of more specific terms which describe exactly what a person is to do to show he knows or understands. Action verbs like these are needed to describe *overt performance which can be observed and measured.*

Describing Conditions and Limitations

In addition to a precise statement of what the student is to be able to do at the completion of training, *the conditions and limitations under which the student will perform must also be described.* If there are conditions which limit or increase the amount of material he must learn when the task is performed under different circumstances, or if the task is easier or more difficult under certain conditions than under others, then those particular conditions must be specified. The conditions that can affect performance generally fall into five categories. First of all, *some practical limit has to be set on the range of knowledge and skill that the student is expected to demonstrate.* For instance, is he to measure any voltage, or only voltages between zero and one hundred? Is the student to name all of the parts of the carburetor, or only the most important ones? Must he be able to operate all of the different kinds of calculators, or only a Smith, Model 5-A?

Many tasks require tools and equipment which may be generally available and standard, or they may be specialized to a particular task. In any event, *the tools needed to perform the task should be listed.* Special job aids may also be provided for many tasks, thus reducing considerably the amount of material to be learned. There is no need for a student to memorize a long procedural sequence if a check list is always available on the job. (It may, in fact, be detrimental to safety and reliability to allow the student to depend on his memory when performing certain critical operations.) Data tables eliminate the need for computing values. Calculators and slide rules can be used instead of computing by hand. Technical manuals and procedural guides contain information to supplement the memory. It is important to know whether special aids will be used during the performance of the task, or whether the task must be performed without their help. Therefore, *training objectives must identify any special job aids that can be used.*

The environmental conditions surrounding the performance of a task are very important. Noise, lighting, temperature, and the accessibility of the equipment being worked on can affect performance. Working in hot, cramped, poorly lighted, uncomfortable situations is very different from working on an idealized bench mock-up situation. If working under difficult conditions is part of the *real job,* then these conditions should be specified in the training objectives. Some tasks have special physical demands; for instance, the task may have to be performed while lying down, stooping, kneeling, or squatting. If any such special physical demands increase the difficulty of the task or limit the length of time the student is able to perform, then the objective should describe those limitations and conditions. Of course, difficult or uncomfortable conditions and limitations should not be imposed during the initial stages of learning, but only after the basic knowledge and skill have been acquired. Eventually, however, the student must demonstrate his skill at the task under realistic on-the-job conditions. *Training objectives must call for performance under realistic job-like conditions to assure a reasonable degree of skill transfer from training to job.*

Setting Performance Standards

The final requirement for a training objective is that it

specify the standards of performance to be met by the student. There are generally two standards, accuracy and time, which taken together are a measure of efficiency. Accuracy standards must be stated in terms of how many of the problems must be done correctly, how many of the items within a question must be answered correctly, and within what tolerances a student must work. *The standards imposed on training performance should reflect the realities of the on-the-job situation.* For instance, there are some cases when performance must be absolutely error-free because of possible danger to personnel, or when one mis-step can lead to serious damage to equipment. However, there are other cases where errors in performance are not critical. Being able to name all the parts in the paper-feed mechanism in a duplicating machine is certainly not as critical a task as reassembling the parts in the correct order. The speed with which the student must perform the task must be specified also. If time is important on the job, if a task has to be completed within a certain time limit, then this standard should be reflected in the objective. There are some tasks for which reasonably lenient time limits can be set for training. However, other tasks may have inflexible time limits imposed by concern for personnel safety or equipment damage. *To be complete, a training objective must clearly state what is realistically acceptable performance.*

Sample Objectives

A well defined training objective must contain the following information: (1) A statement that explicitly describes the *overt behavior by which the learned capability can be observed and measured.* (2) A statement of the *conditions and limitations imposed on the performance.* (3) A statement of the *standards of performance expected of the students.* The sample training objectives listed below have these three components (1, 2, 3) identified to illustrate the amount and kinds of information needed to specify exactly what the student is to do as a direct result of the training.

- The student (1) will match the name of the electronic

components with the symbols (2) on a schematic drawing of any radio circuit. (3) All symbols must be correctly identified. (identifying and classifying)

- The student (1) will list in order all the major moving parts (2) in the power train of a standard shift car. The list will be produced from memory with no assistance. (3) The trainee may omit, have out of sequence, or misname no more than one major part. (verbal chaining)

- The student (1) will start a mobile motor generator, (2) type A-26, using the operating manual as a guide. (3) He must have the generator operating within the specified tolerances (110 ± 3 volts a.c. and 60 ± 3 cycles) within three minutes. (verbal chaining with motor chaining)

- The student (1) will perform simple addition on a calculator, (2) Smith, Model 5-A. He must set up the machine and carry out the addition of sets of five-digit numbers with decimals to three places without the instructions booklet. (3) Ten problems, each with ten five-digit numbers, must be added within five minutes with no procedural or arithmetic errors. (rule-using)

- The student (1) will locate and identify malfunctions in the electrical system of a car. (2) The malfunctions will be shorts and opens induced at logical locations in the system. The student may use a screwdriver, pliers, multimeter, schematic drawings, and the repair manual. (3) Eight of the ten (80 percent) malfunctions must be located and identified correctly within 15 minutes. (rule-using and problem-solving)

Explicitly detailed objectives like those above tremendously simplify the instructional designer's task of determining the types of learning involved and the conditions for learning needed in a training program. For example, if the objective is to start a mobile motor generator, a motor/verbal chain must be learned. If the objective is to match the names of electronic components with their symbols, it is fairly clear that there is discrimination and possibly concept learning involved. If the objective is to list in order all the major moving parts in the drive train, learning a verbal chain of concepts is involved. Performing simple addition on a calculator obviously calls for rule-using along with motor chaining. Having the student locate and identify malfunctions in the electrical system of a car involves rule-using and problem-solving.

Terminal and Interim Objectives

In most cases, several objectives are needed to support each of the tasks delineated in the analysis. The components of a job—duties, tasks, activities, actions—form a hierarchy ranging from very general duties to very specific actions. Similarly, *objectives organize themselves in a hierarchy of increasing specificity, from very general objectives to very specific sub-objectives.* The general objectives are described in more detail by successively more specific subordinate objectives. General objectives closely parallel the duties set out in a task analysis. For instance, the duty of "engine tune-up" converts directly to the general training objective, "learn to perform an engine tune-up." However, this statement of an objective obviously is too general to be of any value except as the starting point in the objective writing process.

The course designer must get down to the task level of description for his information if he is to produce a really useful objective, but *even objectives derived directly from tasks often prove to be too general.* Thus, the designer usually finds himself developing a *series of increasingly specific sub-objectives.* The subordinate objectives are needed because several different knowledges and skills usually must be learned in support of the overall task objective. In learning to perform a certain task, the student may first have to learn to do any or all of the following:

- *To apply general principles* that help to solve unexpected problems that arise.

- *To use specific rules* that apply directly to the task at hand.
- *To follow a procedure* in carrying out the task.
- *To manipulate the tools* required by the task.
- *To name and locate* equipment components and special tools.

Most tasks break down to reveal a similar logical hierarchy of skills and knowledge—proceeding downward from the higher learned capabilities to the simpler behaviors. Note how the structure and sequence of sub-objectives parallels that of the kinds of learning discussed at length in Chapters III and IV. The hierarchy of learned capabilities provides a powerful tool for analyzing and organizing the course objectives. The content, structure, and sequence of the *objectives become identified at the outset with the kinds of learning involved,* and subsequently with the proper conditions for their attainment. Thus, the course designer, at a later stage in the development process, *can derive the instructional strategies directly from the training objectives themselves.*

There are basically two kinds of objectives, terminal and interim. *Terminal objectives state what the student must do to demonstrate mastery of the job* and are derived directly from an overall task. *Interim objectives, on the other hand, are statements of the subordinate skills and knowledge* that must be acquired before the student can master the terminal objective. A set of sample objectives based on the tasks analysis information in the previous chapter will illustrate the hierarchy of terminal and interim objectives.

Statistical Clerk

General Objective – 5

The student must be able to perform simple arithmetic operations accurately and quickly on an automatic calculator.

Terminal Objective – 5.1

The student will perform addition of decimal numbers on a Smith Calculator, Model 5-A. He must set up the machine and carry out the addition of five-digit decimal numbers to three places without the aid of the instructions booklet. Ten problems, each with ten five-digit numbers, must be added in less than five minutes without any procedural or arithmetic errors.

Interim Objective – 5.1.1

The student, before using the calculator, will demonstrate proficiency in adding five-digit, three-place decimals by completing test sheet No. 19. He must finish the test in less than 15 minutes and solve at least 85 percent correctly without outside help. If he does not succeed, he will be given remedial self-study materials and tutorial help to bring him up to proficiency.

Interim Objective – 5.1.2

The student will demonstrate to the instructor's satisfaction the correct way to set up Auto-Decimal for five-digit, three-place decimal numbers.

Interim Objective – 5.1.3

The student will demonstrate to the satisfaction of the instructor the procedure for clearing the keyboard in preparing to operate the machine. If the keyboard does not clear, he will demonstrate the alternate procedure. If the keyboard clears properly the first time, he will explain the alternate procedure for clearing.

Interim Objective – 5.1.4

The student will identify and locate all the function keys and controls on the keyboard by writing their names in the blanks provided on the special cardboard mock-up. All 22 function keys must be named from memory in less than five minutes with only three errors.

The examples above demonstrate how objectives proceed downward from the general to the more concrete and the more specific. The samples also demonstrate how the tasks involved lend themselves to an organization based on the hierarchy of learned capabilities, proceeding downward from a complex rule-using task to the simple task of identifying.

In this case, four interim objectives were needed to specify all the intermediate learned capabilities needed to support the one terminal objective. Each of the interim objectives must be reached before proceeding to the next, and all must be reached before proceeding with the terminal objective. However, regardless of the number of intermediate skills and knowledges required, the final pay-off is the capability called for in the terminal objective. *The student must ultimately be judged on his performance of task-level terminal objectives, not on the separate activity-level interim objectives.* (See Appendix A for further samples.)

Instruction is basically the management of the conditions of learning. To be effective and efficient, a training system must bring about the greatest amount of change in performance capability in the shortest possible time. The key decisions during the instructional systems development process are concerned with the planning of suitable conditions for the specific types of learning. *Clearly stated and explicity detailed behavioral objectives are the essential first step toward sound instructional design.*

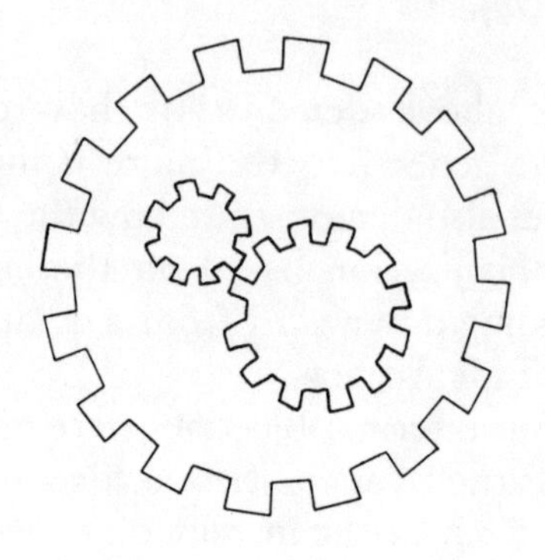

CHAPTER IX

THE CRITERION TEST

Introduction

Assessment of student performance in relation to the stated training objectives is critical to an instructional system. Actually, five kinds of assessment are needed.

1. Before-lesson assessment that is diagnostic, prescriptive, and directive in nature; thus allowing the student to concentrate on areas of weakness or perhaps to by-pass the lesson entirely.
2. Immediate and continuous within-lesson assessment to furnish the student the feedback that is an integral part of the learning process itself.
3. Immediate and continuous within-lesson assessment to confirm attainment of each capability before proceeding to the next, because each learning experience systematically builds on a preceding learned capability.
4. End-of-lesson and end-of-unit assessment to predict the capacity of students to proceed to related or advanced lessons and units.
5. End-of-course assessment to predict transfer of knowledge and skill to on-the-job situations, and to predict performance in related or more advanced courses.

All five kinds of assessment are very important and must be built into the instructional system from the very beginning. However, there is another requirement for assessment that is perhaps even more important, because it is the very foundation of the systems development process—the validation of the system itself by pre-testing it against the stated objectives. *The validation process, during which the training system is brought up to a specified level of effectiveness by a "produce-test, revise-retest" development cycle, is a major innovation in training course design.* The vehicle for validating the system is the criterion test; thus, next to the training objectives, it is probably the most important design document produced during the systems development process. *The concept of using a criterion test to measure the effectiveness of the instruction rather than to measure student proficiency is truly a breakthrough in education and training methodology.*

The criterion tests serve as quality control instruments by comparing the performance demands placed on the student during training with those of the actual job for which it is preparing him. To do this, the *criterion tests are used to evaluate the training objectives, to evaluate individual lessons, to evaluate the complete system, and to continuously evaluate the system after it has been implemented in the classroom.*

Group test data gained from sampling performance on the criterion tests are used two ways: first, to evaluate overall system performance and, second, to locate and diagnose trouble spots in the system. (Analysis of an individual student's performance on a criterion test can quickly locate his trouble spots, too.) How the validation steps are carried out is not the concern of this chapter. The details of the validation process will be covered in later chapters. The discussion here will mainly cover the development of the test itself.

Characteristics of a Criterion-Referenced Test

To be a true measure of the student's learned capabilities, *the criterion test must be criterion-referenced, comprehensive, valid, reliable, objective, standardized, and economical.* Of course, no single test can be perfect in all respects because, as always, some of the desirable characteristics will have to be traded off for others. Most course developers and instructors are familiar with the above terms, but when used in relation to a criterion test these terms take on somewhat different and special meanings. The differences all stem from the fact that this is a criterion-referenced test rather than a norm-referenced test.

Training under the instructional systems concept is ideally "learning-to-mastery" and "training-to-proficiency"; that is, every student is to achieve at least the minimum standards set forth in the objectives. The student always has to achieve a specified level of proficiency in the task at hand before being allowed to move on to the next. *The individual's proficiency is measured against a predetermined set of absolute criteria, rather than relative to the performance of the other students.* Norm-referenced tests that rate individual performance against group norms cannot be used to evaluate either the students or the training, when there are explicitly stated performance objectives. Heretofore, test construction methods have concentrated primarily on techniques that were designed to discriminate among students and to accentuate their differences by rating their performance along the so-called normal curve. *In an instructional system the goal is not to differentiate among the students, but to raise every individual's performance to the level specified as acceptable by the objectives.* Thus, the main purpose of the criterion-referenced test is to determine as accurately as possible *when* a student has reached the acceptable level of performance. Criterion-referenced instruction allows latitude in the amount of time taken to reach proficiency, but allows no latitude below a minimum level of proficiency. There is a fundamental difference between criterion-referenced tests and norm-referenced tests—*criterion-referenced testing separates the students along a time scale, while norm-referenced testing separates the students along a proficiency scale.*

Actually, criterion-referenced training does allow for some individual differences in performance, but the range of difference is much more restricted than in norm-referenced training. Though all students must eventually meet the minimum standards (which may be quite high), there is still room at the top for some sort of rank ordering of the individuals. Morale and motivation may still need incentives based at least to some degree on relative performance. Because of the restricted ranges of performance measured by a criterion-referenced test, it is probably best to use only a pass-fail or a superior-satisfactory-incomplete grading system. The student whose performance surpasses the minimum standards set forth in the objective merits the "superior" rating. The student who only meets the minimum standards would be rated "satisfactory." The "incomplete" is usually only a temporary indication of an unsuccessful attempt to meet the minimum standards. Ideally, every student should eventually perform at the 100 percent level—achieving the minimum standard specified by the objective is 100 percent. Realistically, however, there will always be a few students who cannot achieve some of the objectives despite repeated efforts. In such cases their record should reflect the missing capabilities. A recommended format for an "Occupational Readiness Record" is shown in Appendix L.

If, however, some sort of ranking is needed to indicate relative performance, *only the major end-of-unit review tests, such as a six-week review, should be given percentage or point grades.* Any such grade must not be taken as a measure of past performance, but rather as an indication of what has been retained and thus available for transfer to new situations. All students have had to meet at least the minimum performance standards as they have progressed from lesson to lesson. Therefore, *relative ranking must be considered as merely a means for predicting a student's ability to go on to more advanced work.*

The items in the criterion test are derived directly from the previously developed training objectives by recasting them as questions or requests for demonstrations of specific skills and knowledge. Therefore, if the training objectives have been prepared properly, the task of writing the criterion test should be relatively simple. *Good training objectives furnish all the necessary information for good test items*—the specific capability to be measured, the conditions and limitations under which the student will perform, and the level of proficiency to be achieved. Usually, for each objective, both terminal and interim, a companion item must be included in the criterion test. Because the objectives are already criterion-referenced, the items derived from them will be also. With an item for every objective, the test will necessarily give comprehensive coverage of all desired behaviors. And, if the objectives have job validity, then the test will also have true validity. Likewise, if the objectives call for behavior that is observable and measurable, the test will probably have a high degree of reliability. *If a thorough and accurate task analysis were conducted and the subsequent training objectives were carefully developed, the criterion test would, in all probability, be criterion-referenced, comprehensive, valid, and reliable.* There will, of course, be errors of omission and commission during each phase of the development process preceding the writing of the criterion test, and so there is still a need to validate the test. The procedure for validating the test will be covered in the next chapter.

A good test must be relatively objective; that is, the judgment of the scorer should enter the scoring process as little as possible. Everything else being equal, an objectively scored test which does not permit scorer bias to affect the score is more valid and reliable than a subjectively scored test. Criterion testing should treat everyone the same. As nearly as possible, every student should have the same opportunity to perform under the same conditions and to the same standards of proficiency. The directions, equipment, conditions, limitations, and standards should be the same for every student. *A test is reliable and valid only insofar as it maintains equal treatment for every student.*

The criterion test must also be economical regarding time, manpower, and facilities; but economy is strictly relative when applied to an instructional system. *The gain in reliability and validity through using a truly comprehensive, criterion-referenced PERFORMANCE test far outweighs the economy of group paper-and-pencil tests.* In addition, as emphasized earlier, testing in an instructional system is used for many reasons other than discriminating among the students. Within-lesson testing provides feedback for the learning process and the confirmation of the correctness of performance needed by the student before proceeding to the next step. End-of-lesson testing is needed before proceeding to the next or related lessons, and end-of-course testing is needed to predict transfer. Moreover, continuous assessment of the system itself is needed. Thus, a good portion of the students' time is taken up with testing; but it is not just testing for the sake of testing *because testing is an integral part of the dynamic nature of an instructional system.* Comprehensive performance tests are even more important during the developmental validation of the system. Sample student performance on every objective, both terminal and interim, is needed to provide complete data for validation. *During validation testing, the quality of the data, not the economy of administration, is the main concern.*

In a completed and installed training system, *by far the greatest portion of testing time should be confined to student self-testing*; especially testing for achievement of the interim objectives. Because an instructional system requires almost continuous assessment of progress with concurrent feedback to the students, instructors could be overwhelmed with the administrative and clerical work involved unless self-testing procedures are relied upon. Only in this way can the instructor have the time needed to work with all of his students individually. In addition, giving the students themselves the responsibility for their own day-to-day progress checks contributes to their motivation and positive attitudes toward learning.

Performance on interim objectives, unless very critical, is usually not tested formally. Formal testing generally should be concerned only with the achievement of terminal objectives. Of course, *end-of-unit and end-of-course tests have to be limited to terminal behaviors, and probably only key terminal objectives at that.* Therefore, the items that cover interim objectives are dropped from the formal criterion tests when the system is finally installed in the classroom. Once the system has been validated, there is no further need to include formal testing of interim objectives for diagnostic purposes; thus, formal testing time can be cut considerably in the finished system. Generally, no more than

10 percent of the total course time should be given to formal graded testing. However, the amount of time devoted to informal self-testing can and should run considerably higher. If the students are not spending at least half their time responding and performing, with some sort of self-checking involved, the training program does not require enough active participation. Moreover, some sort of self-check on the achievement of the interim objectives should be built into the material to aid the students in self-diagnosis and prescription for trouble spots.

Within-lesson self-checks are not tests in the usual sense; instead they are means of providing meaningful feedback to the students. The internal self-tests thus become an integral component of the learning process itself and, as such, should never be scored as part of a cumulative grading system; but only to aid the student in self-diagnosis and to help the instructor in assessing system performance. The students must feel free to make errors without fear of punishment if the self-tests are to function as aids to learning. Under these conditions, the students will come to realize that cheating or rationalizing away errors on self-checks is self-defeating and that discovering one's own errors is merely the first step towards "learning-to-mastery." (See Appendix G for a sample self-scoring test device.) Moreover, giving the student actual control over this portion of the learning process encourages the development of self-discipline and a sense of responsibility for his own progress and achievement. It cannot be overemphasized that the *self-testing procedures must be treated as central to the learning process, and therefore must be completely non-aversive.*

Types of Tests

Objective paper-and-pencil tests are usually printed and have the directions included right on the form. The student either writes his answer on the test itself or on a special answer sheet that comes with the test. Basically, these are knowledge tests, with the questions in the form of multiple-choice, completion, or matching exercises. The questions may be supported with numbers, diagrams, pictures, or any other material that can be printed. Most commmercial aptitude, achievement, or intelligence tests come in this form. The principle advantage of this type of test is objectivity in scoring, for it can be scored quickly and accurately by almost anyone. There are also special answer sheets that can be scored by machine. The number of individuals that can be tested at one time is limitless. Tests such as these can be given to one or a thousand at a time and may also be self-administered and self-scored. *The advantage of objectivity and economy make such tests very popular*; however, they also have their disadvantages. It is difficult to imagine a paper-and-pencil test that could possibly measure the behaviors involved in operating heavy earth-moving equipment. Just because a student can list the steps for starting the diesel engine and can explain how to raise and lower the blade on earth-moving equipment is no guarantee that he can actually perform the operations. Of course, there are paper-and-pencil tasks such as those concerned with reading, writing, and computing that can best be measured in such tests, but *there are many behaviors that cannot be measured validly with paper-and-pencil tests.* The principal use of objective paper-and-pencil tests is to check on the acquisition of the specific supporting knowledge content of interim objectives.

Essay tests generally present a broad question and ask that it be answered at length. The examinee is asked to "write everything he knows" about the question. Essay tests are relatively easy to prepare but extremely difficult to score. The subjective opinion of the scorer is a heavy factor in determining the score. In addition to the lack of objectivity, the only valid reason to use an essay test is to measure the students' ability to write. *Essay questions should not be used on the criterion test unless the training objectives call for a demonstration of essay writing ability.*

Oral tests are most often given the same way essay tests are conducted, except the student must talk rather than write. The instructor asks the student to "tell all he knows" about something. Under these conditions, oral tests suffer from exactly the same difficulties and faults of essay tests. Additionally, an oral test can only be administered to one student at a time. However, oral tests do not have to be built around general questions. In the form of a carefully controlled interview, questions can be asked which require short, straight-forward answers. When short-answer questions are used along with a standardized objective scoring system, oral tests are acceptable as criterion tests. In fact, there are many instances where the performance of certain tasks can best be checked by asking the student to "show and tell." *Having the student demonstrate and explain the steps in a complex procedure is often the most logical, practical, and efficient way to test competency.*

Performance or skill tests require the student to perform specific tasks, rather than merely supplying information as in the typical knowledge test. Usually the tasks are samples of work associated with a certain job. An obvious performance test for a typist would be to require the student to use a typewriter to copy a passage. The performance would then be scored in terms of accuracy and the time needed to complete the job—words per minute correctly typed. Usually, a skill test requires the students to use the tools and equipment they will use on the job. There are also times when special or simulated tools and equipment created just for training and testing purposes are used. Performance tests generally have to be administered to one student at a time and they frequently require considerable time. They are also expensive in terms of manpower, space, and equipment, and only rarely can they be self-administered and self-scored. In spite of these disadvantages, *performance tests are strongly recommended as criteria for vocational and technical training programs.* In fact, if the course designer follows the prescribed steps in arriving at truly behavioral objectives and then derives the criterion test directly from those objectives, the test will, of necessity, emphasize job-like performance.

Because test construction is a highly specialized subject, a large part of the Appendices (Appendices B through F) is devoted to a rather detailed discussion of some recommended techniques. Good test construction is absolutely essential to the success of instructional systems development. *The entire developmental process can stand or fall on the quality of the criterion tests.*

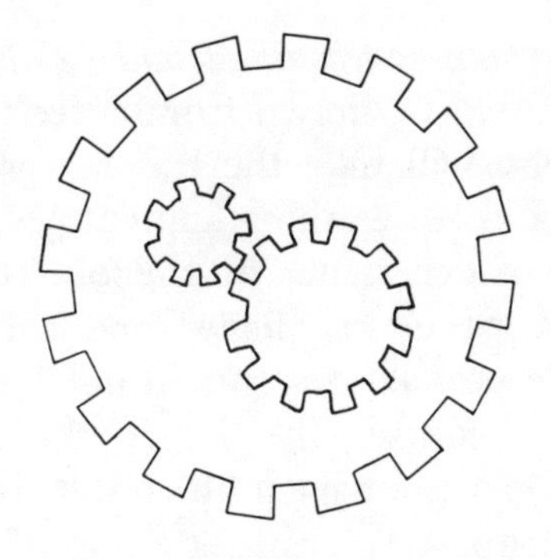

CHAPTER X

VALIDATING SYSTEM CONTENT

Introduction

During the preceding stages in the systems development process, the emphasis has been on the collection of information, not on the evaluation of that information. Conducting a task analysis, formulating training objectives, and constructing a criterion test are systematic processes, of course, but the evaluative decisions are quite subjective and mostly the personal judgment of the course designers. This is not to say that the designers' judgments as to what to include in a course are not valuable and necessary as a beginning. However, observation, interview, expertise, and historical precedent can go only so far—they can only set the preliminary pattern. Thus, the instructional systems concept calls for objective, empirical evaluation of the proposed content before proceeding further with course design. *The criterion test provides the means for carrying out an objective validation of proposed course content.*

Validating the Criterion Test

Before the criterion test can be used to validate course content, the test items themselves must be validated. To do this, *the test is given a preliminary tryout on two different sample populations: (1) untrained-unskilled and (2) trained-skilled.* The unskilled sample should be drawn from entering students who are typical of those who will take the training program. The skilled sample should come from working apprentices who have approximately six months of experience on the job. The apprentices may be recent graduates or they may have come from entirely different sources. The sample populations should number at least 30 to 50 in each group. Of course, the larger the sample, the more confidence the designer can have in the results of the validation.

If time and manpower permit, a group of 30 to 50 students who have just completed an on-going training program can also be tested. This additional group will add to the range and type of data available for analysis. *Only the data from the trained-skilled sample are used to analyze and to improve test validity, but the data from both the trained and untrained samples are used to analyze and to improve test reliability.*

There are several reasons for giving the criterion test a preliminary tryout with sample trained and untrained populations.

- To iron out the administrative details involved such as the scheduling of facilities and instructors, availability of equipment and materials, etc.
- To try out procedures for giving the tests such as directions, time limits, scoring, weighting, etc.
- To allow the instructors to gain competency in performance testing which often calls for observing and evaluating individual performance of specific tasks.
- To check on the validity of the items by analyzing the response data from the trained population to determine if the items actually evoke the performance or behavior they are supposed to evoke; and to revise the items as necessary—thus establishing the validity of the test.
- To check on the reliability of the items by analyzing the response data from both the trained and the untrained populations; and to revise the items as needed to eliminate ambiguities, errors, misleading wording, cues to the right answer, poor alternatives in multiple-choice

questions, etc., each of which may prevent the items from accurately and consistently measuring the specified performance or behavior—thus establishing test reliability.

Analysis of the response data will answer most questions about the adequacy of the criterion test as a testing instrument and will provide the information needed to revise items. The familiar concepts of validity and reliability actually take on new and broader meanings in a criterion-referenced test. There is no longer a need for complex statistical procedures; rather the process becomes a simple, straight-forward, common-sense analysis of actual performance in relation to desired performance. *A criterion-referenced test is valid when it is composed of items that evoke or stimulate the desired behavior in at least 85 percent of a sample trained population. A criterion-referenced test is reliable when it is composed of items that accurately measure the desired behavior in 85 percent of a sample trained population.*

To determine validity and reliability, a kind of item analysis is performed, but it must be emphasized that this is not the usual item analysis associated with regular test and measurement concepts. As pointed out in the preceding chapter, the construction of criterion-referenced tests is vastly different from that of norm-referenced tests. Items that are changed or eliminated as undesirable in a norm-referenced test are highly desirable in a criterion-referenced test; moreover, the items are kept in a criterion test *for the very reason* they are dropped from a norm-referenced test. For example, when constructing a norm-referenced test, considerable care is taken to eliminate most items that are too easy—if many over 50 percent get the item correct during tryout, the item is considered too easy; so the item is thrown out or changed to make it more difficult. In direct contrast, if an item on a criterion test is not responded to correctly by at least 85 percent of the sample trained population, the training is considered inadequate—rather than the item being too easy.

The goal of a training system is to have each item on the

criterion tests responded to correctly by at least 85 percent of the students. Thus, a test item is never changed arbitrarily to make it less or more difficult. *The basic content of each test item is derived directly from a specific training objective, and so the difficulty level cannot be changed without changing the objective.* The structure or wording of the item, however, may have to be changed to make it a more accurate measure of the desired performance. Careful analysis of the items that less than 85 percent of the sample get right is needed to determine if the item is poor or if the training was inadequate. Actually, the course designer is interested only in correcting weaknesses in the items at this point. Correcting weaknesses in the training comes later during the test, revise, and retest cycle, when the training materials are being produced and evaluated.

Item analysis for norm-referenced tests usually includes determining whether the items discriminate between the "good" and "poor," "bright" and "dull" students. When as many poor as good students get an item right on a norm-referenced test, that item is said not to discriminate; and when more poor than good students get an item right, that item discriminates in the wrong direction. In either case, the item is dropped from the test, or it is changed so it will discriminate correctly between good and poor students.

As discussed in the preceding chapter, the purpose of a criterion test is not to discriminate among the students, but to determine instead *when* each student has reached the acceptable level of performance specified by the objectives. It might be said there are no "good" and "poor" students under the instructional systems concept, only slower and faster. And, *because the goal of a training system is to have each objective achieved by at least 85 percent of the students, the items on a criterion test obviously cannot discriminate very well between the "good" and the "poor" students.* However, if an item in a criterion test does discriminate—that is, most "good students" get it right and most "poor students" get it wrong—the training may be at fault, not the item. Of course, items can be confusing, ambiguous, and misleading; and so, for the wrong reasons, they may not adequately measure performance. Perhaps there are built-in cues that give away the answer, or maybe the item is so confusing that no one can respond correctly. Careful analysis of each item that fails to reach 85 percent is needed to determine whether the *training* or the *wording of the item* is at fault.

Although a criterion test does not discriminate very much among students, *a criterion test should definitely discriminate between the trained and the untrained sample populations.* If a majority of the untrained population or only a few of the trained population (both the apprentices and the recent graduates) respond correctly to an item, there obviously is something wrong with the item. The item has to be revised to eliminate the cues that give away the answer to the untrained or the errors that mislead the trained. No test item can be eliminated, of course, because each is needed to measure behavior required by a specific training objective. Suppose, however, there are no structural problems with the item; it is a clearly and properly stated item, and yet the majority of the untrained get it correct. What does this mean? Or, what does it mean when the majority of the trained sample cannot respond correctly to an item? These questions lead to the next section which deals with content validity.

Validating the Training Objectives

After the item analysis of the criterion test has been completed and the items have been revised to bring the validity and reliability to a satisfactory level, the data from the sample skilled and unskilled populations are next used to validate the training objectives. Actually, both processes, test validation and the validation of the objectives, can be carried out at the same time. The designer must, however, be very certain he has eliminated the possibility of the item itself being at fault before he questions the validity of the course content.

Two key questions have to be asked about every training objective, both interim and terminal, when validating course content:

- Does the objective specify behavior that is not already

within the capabilities of the majority of the untrained sample?

- Does the objective specify behavior or performance that is actually needed by the majority of the entry-level workers who have been on the job for about six months?

The answers to these two questions come from further analysis of the data derived from giving the criterion test to sample populations of the unskilled and the skilled. *The untrained-unskilled sample must be typical of the students about to enter that particular course. The trained worker must have about six months on the job* for which the course is specifically being created. It is important that apprentices are sampled because the goal of the training is to turn out good entry-level workers, not skilled journeymen. In addition, it is important that the sample apprentices have only about six months of experience on the job because by that time most of them will have forgotten the needless knowledge and skills they were exposed to in training, and most of them will not as yet have acquired much of the advanced knowledge and skills needed by the skilled journeymen. Also, after about six months on the job, most of them should be highly experienced in their entry-level tasks, the goal toward which the training program should be aimed.

Assuming the test item is valid and reliable, what does it mean if the majority of the *untrained* population responds to that item correctly? *When 85 percent or more of the untrained population responds to an item correctly, it means the item is asking for behavior that probably does not need to be taught.* Based on the sample tested, the course designer can predict that most of the entering students will probably be able to achieve the objective, from which the test item was derived, without benefit of the proposed training. If most of the similar or related test items from the same task cluster (the terminal and interim tasks) are also performed correctly by 85 percent of the sample, then there can be no doubt that those training objectives (both interim and terminal) are not valid content for the course. When over 85 percent of the target population can already meet those objectives without the training, there is obviously no need to include them in the course content. *Performance of the sample untrained population on the criterion test determines the level of knowledge and skill that most of the entering trainees will bring with them, and thus analysis of the test data sets the lower limit of the course content.*

In contrast to the above situation, what does it mean if the majority of the sample *trained* population fails to respond correctly to a valid and reliable item? *Failure of 85 percent or more of the trained apprentices to perform correctly means the item is asking for behavior that is probably not needed for the job.* If the successful apprentices (and they are assumed to be so after six months on the job) can do the job well without that capability, then apparently there is no need for it. Failure to perform similar and related items from the same task cluster confirms that the training objectives from which the items were derived are not valid course content. The findings can be further substantiated if the results of testing a group of just-graduated students show that most of them can respond correctly to items that the graduates who have been on the job for six months can no longer perform. Of course, there may be instances where certain objectives would be retained despite such findings, because expert opinion may foresee their value. However, at the very least, such negative findings should raise serious questions concerning the validity of the objectives. *Performance of the sample trained population determines what skills and knowledge are actually needed on the job by the entry-level worker, and thus analysis of the test data sets the upper limit of the course content.*

Using sample untrained and trained populations to validate both the criterion test and the course content is a vital step in the development of an instructional system. The 85 percent figure is pivotal in the analysis of performance data because it includes all but one standard deviation of a normally distributed population as defined by the normal curve. Whenever the data for a population of 30 or more students fall near 85 percent, the course designer has a firm base on which to make his decisions. However, even when the data range between 50 and 85 percent (better than chance) he should be alert for possible trends and indications of problems. *Empirical validation of course objectives against the actual behavioral demands of the job itself is a major breakthrough in training methodology.* However, this is only the first step in the continuing validation process that is built into an instructional system. The validated criterion test will be used to assess the effectiveness of the system as it is being developed, and it will be used to evaluate system performance as long as the system is in use. Analysis of performance data derived from the criterion test furnishes the information that is the basis for maintaining the quality of the training and for revising the content of the training. *The validated criterion test thus serves as an agent of both stability and change.*

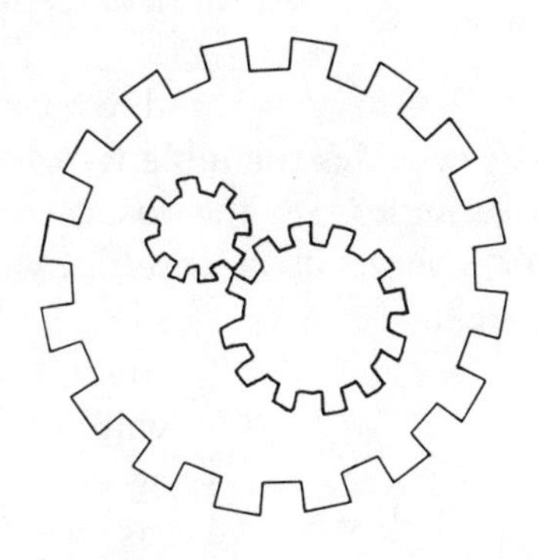

CHAPTER XI

ORGANIZING SYSTEM CONTENT

Introduction

Up to this point in the development process, each step has been directed toward constructing a strong framework on which to build. The validated training objectives form the skeleton of the training system; course content is the flesh and muscle. The supporting knowledge and skills that are essential to performance must first be organized before developing the actual course materials. Course content has to be structured and sequenced in the best possible order for learning, if the system is to be effective and efficient. This chapter will describe a method for developing a logical course outline.

Content sequence is based on a combination of two different levels of organization—a general sequence based on job/task structure and a more specific sequence based on learning structure. As mentioned in the earlier chapter on learning, having the correct sequence of learning events is extremely important; although there is probably no one best sequence. The method advocated for arriving at a sequence will produce a very satisfactory preliminary structure, but it should be remembered that this structure is only preliminary and thus may not necessarily be final. *Analysis of student performance data during the "test-revise-retest" cycle will determine final sequences.*

Job/Task Sequence

The task analysis provides the designer with an outline of the duties, tasks, and activities in the order in which they are performed on the job. The structure could be visualized something like that in the chart below:

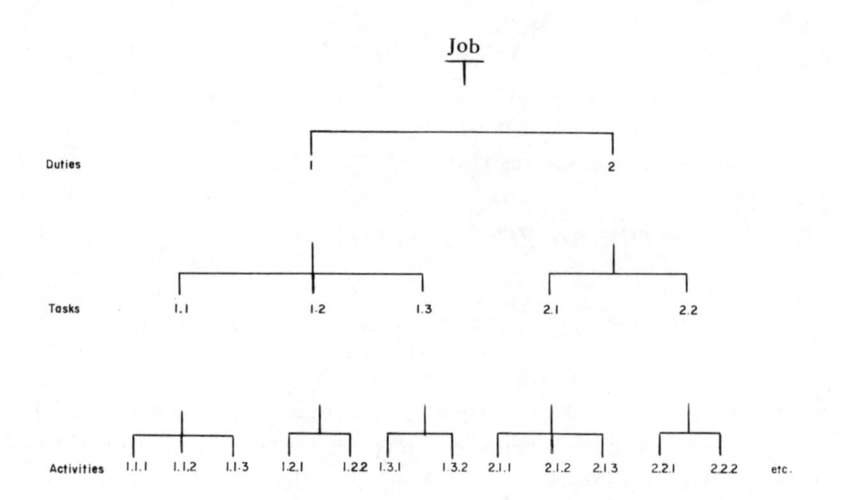

The sequence of the on-the-job activities should be used as a guide to the overall sequence of learning events. *In general, the learning of the activities, tasks, and duties should follow the sequence set by the job itself*; that is, activity 1.1.1 would be learned before activity 1.1.2, and 1.1.2 before 1.1.3; task 1.1 would be learned before task 1.2, and 1.2 before 1.3; duty 1 would be learned before duty 2, and 2 before 3, etc. Thus, the general structure of the course can be derived directly from the task analysis documents.

Job-sequenced training is basically a sound approach to the problem of course structure. *Acquiring the job elements in the order they will be performed on the job lends a great deal of realism to training,* heightens the transfer of knowledge and skill from training to job, organizes the course in a pattern that is logical and understandable to the student, and gives the student clearly visible goals with an accompanying sense of real progress. Rigid following of job sequence is, of course, not desirable. There are instances when job-sequenced training may not be practical, efficient, nor effective. However, it is probably best to be guided by the on-the-job sequence of events whenever possible.

Learning Structure

The second phase in preparing the content outline is to organize the supporting knowledges and skills. Each activity, task, and duty usually requires supporting knowledge and skills. As indicated in the earlier chapter on learning, *the different learned capabilities require different prerequisite conditions; so there has to be a careful sequencing of the internal learning events within each of the main job-sequenced events.* In general, the learning sequences can be plotted by "working backward" from the training objective to determine each subsequent prerequisite capability—from a higher-order principle back to a supporting sub-principle, from the sub-principle to supporting concepts, from one of the supporting concepts to a sub-concept, etc. Careful analysis of the supporting knowledges and skills (working backward from the objective task) will produce a logical learning structure. This can be a rather demanding and time consuming job; however, a technique of matrix analysis can help the course designer considerably with this complex job. *Matrix analysis depends on both the associations and the discriminations among the related supporting skills and knowledges as the basis for organizing the learning structure.*

The first step in producing a matrix is to prepare a list of terms, concepts, rules, and principles needed to support each objective. Each concept or rule should be written as a simple statement of fact. No statement should contain more than one main idea. If the idea or concept is very complex, each statement should cover only one aspect of the total idea. These statements of fact may be listed in a somewhat random order at first. There is no need to attempt more than a common grouping of the facts at this time. The real analysis must await later steps.

The second step is to number each statement of fact for reference.

The third step is to prepare a graph, as shown in Figure 1, labeling the squares as required to represent each of the facts. Place the numbers diagonally on the graph from the top left to the bottom right.

The next step is to decide which facts are associated with which other facts on the matrix and to indicate these associations by marking the common square on the graph as in Figure 2. Notice that some facts will be associated with several other facts; for instance, fact No. 1 is associated with No. 3 and No. 4, and fact No. 3 in turn is associated with both No. 1 and No. 2. Each fact must be checked against every other one on the matrix. Be sure to indicate all associations.

Indicate an association between two facts or rules only if they have common elements. In a given chain of events, the first and last events are not directly associated; rather, the connection between them is established indirectly by the intervening related facts in the total chain.

The next step is to look for the important differences between the concepts and the rules. Discriminations are essential to the understanding of any subject. What something is not (or does not do) is as important as what something is (or does). Lesson designers should take advantage of possible discriminations when developing learning experiences. Through discriminations, the student can get a better total grasp of a concept or rule; he is less likely to forget it, and he will be able to apply it over a broader range of new experiences.

Check each fact against every other one on the matrix and note any important differences. Indicate the discriminations on the matrix by marking in the common squares below the diagonal. Figure 3 shows a completely analyzed matrix with both the associations and discriminations marked in.

The last step is to make a new matrix with the concepts and rules resequenced. The goal is a sequence that will put the associations as close as possible to the diagonal listing of the rules. By moving the related facts (those with marked associations) as

Figure 1

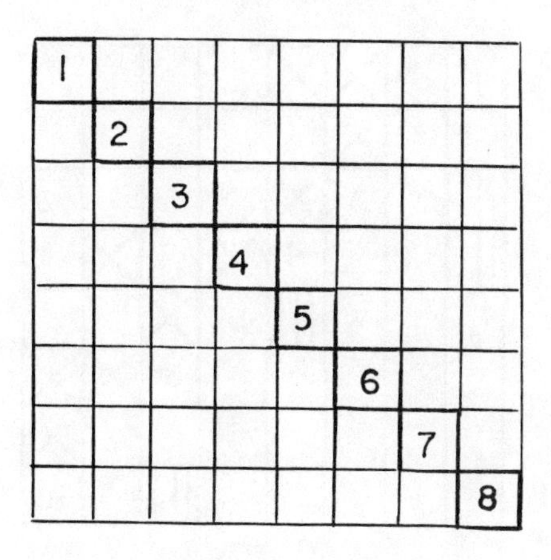

Figure 2

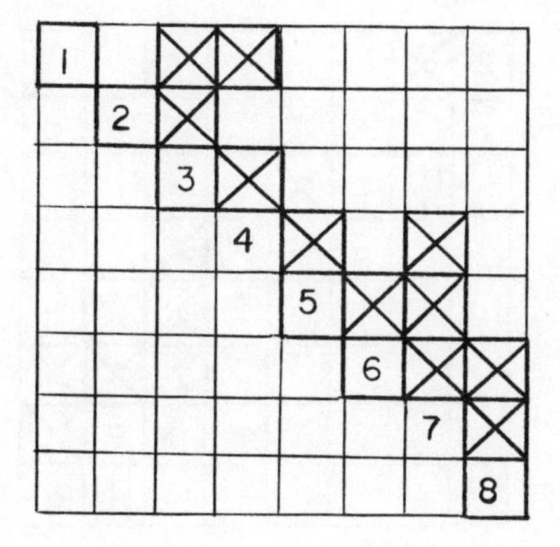

Figure 3

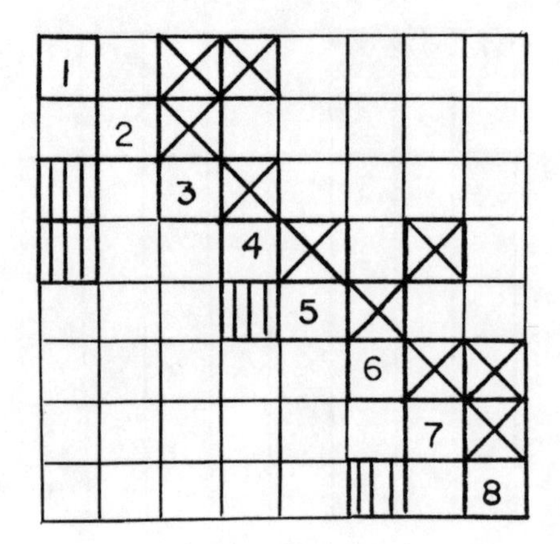

Figure 4

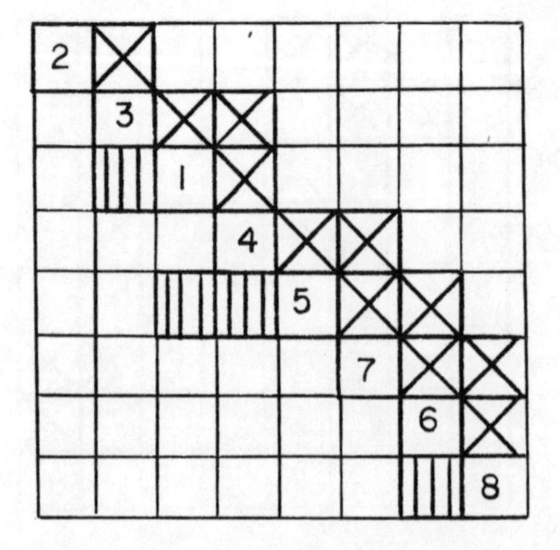

close together as possible, the associations will cluster next to the diagonal. The process may take considerable shuffling and reshuffling of the facts to get a good fit. There are often several good fits, leaving the designer to choose the one he feels is the best compromise. The rearranged matrix is shown in Figure 4.

The discriminations should also be close to the diagonal line, thus insuring that the important discriminations will be taught along with the associations.

For any given lesson, there may be several acceptable sequences. *The best sequence will plot the closest grouping about the main diagonal of concepts and rules.* Compare Figures 3 and 4 and note the improvement in the grouping of the common elements.

To illustrate what is involved in performing a matrix analysis, a sample lesson in basic electricity is analyzed below. The development of this matrix will demonstrate the process in more detail.

The following concepts and rules form the basis of a lesson on Ohm's Law:

1. Current is the flow of electrons.
2. The rate of flow of electrons in a circuit is stated in amperes.
3. Ohms are related to volts and amperes by Ohm's Law ($E = IR$).
4. The symbol for voltage is E.
5. The symbol for current is I.
6. The symbol for resistance is R.
7. The quantity of electrons in a circuit is stated in terms of the coulomb.
8. An electric charge is produced by friction.
9. The basic unit of charge is the electron.
10. The ampere is defined as one coulomb per second.
11. The volt is the unit of electromotive force.
12. Electromotive force is the rate of supply of energy per unit of current.
13. Energy, or the capacity to do work, is measured in

joules.

14. The volt may be expressed as joules per coulomb.
15. Current meets resistance in flowing through a load.
16. Work is done in overcoming resistance.
17. The pure resistance in a circuit is measured in ohms.
18. The potential drop or pressure drop from one point to another in an electrical circuit is measured in volts.
19. Voltage, electromotive force, and electrical pressure are synonymous.
20. Current flow, electron flow, and current are synonymous.
21. The two requirements for electron flow are a complete circuit and a source of voltage.
22. The three forms of Ohm's Law are $E = IR$, $I = \frac{E}{R}$, and $R = \frac{E}{I}$.
23. Given any two values of E, I, or R in a simple d.c. circuit, you can solve for the unknown quantity by using the appropriate form of Ohm's Law.
24. An electrical circuit consists of an electrical source, a conductor, and a load.
25. Voltage is what makes electrons move.
26. Current is electrons in motion.
27. Resistance is the opposition to a flow of electrons that reduces the amount of current.
28. An electrical circuit is a closed path for electron flow.
29. One ampere is the amount of current produced by one volt of potential difference across one ohm of resistance.

First, the rules are numbered and entered on the graph as shown in Figure 5.

Next, a detailed analysis is made and the associations and discriminations noted on the matrix. From this a new matrix is constructed with an improved sequence. The final matrix is shown in Figure 6.

It should be noted that in the process of analysis, any

MATRIX ANALYSIS

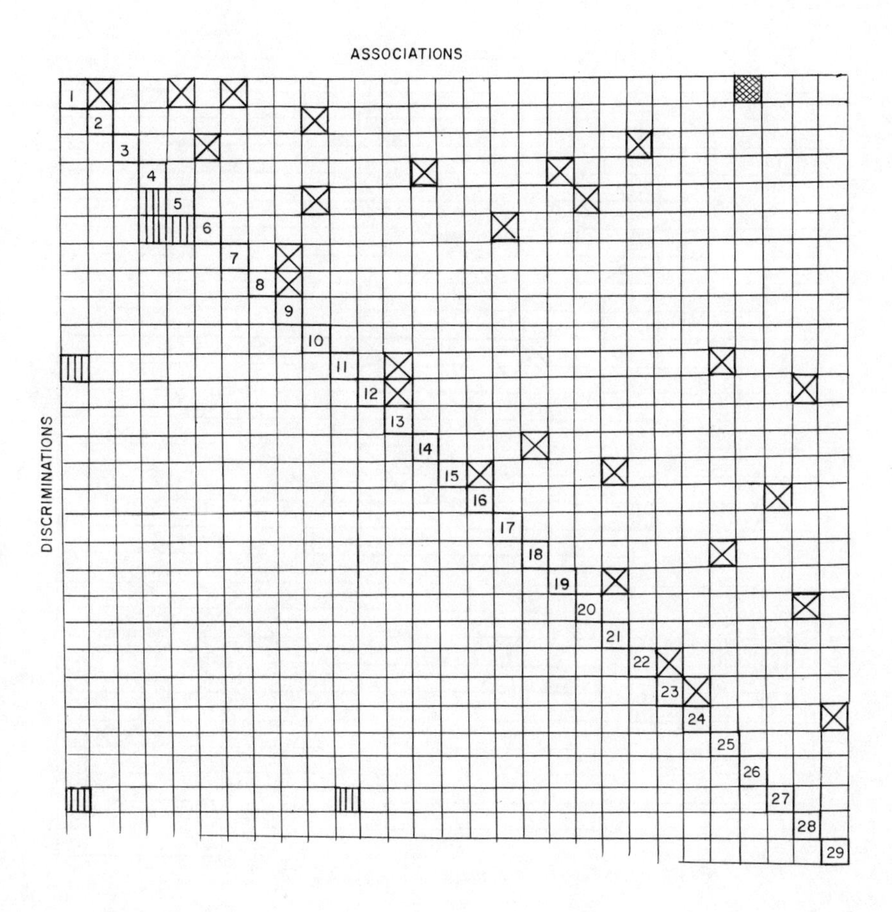

Figure 5

Figure 6

MATRIX ANALYSIS

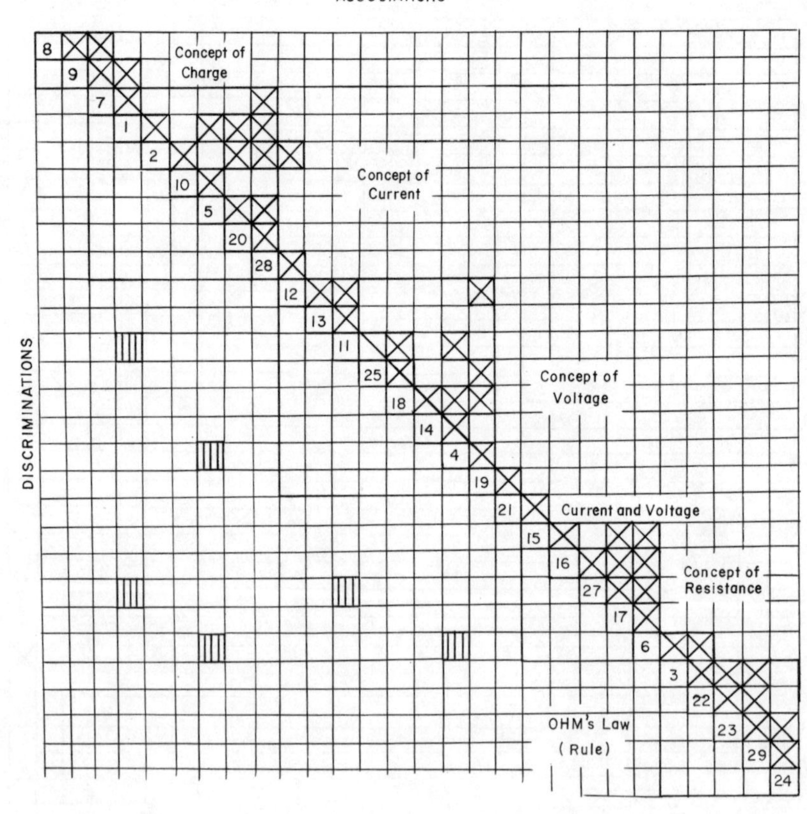

duplication in basic facts or rules will also be uncovered. For instance, rule 26 on the original list is just another way of stating rule 1. Thus, in the final matrix, rule 26 has been eliminated.

Notice that five concepts will be covered in the final lesson arrangement. Within each concept area there is a close and natural grouping of the related facts. Between every adjacent fact there is at least one association and often there are several associations with other facts in the concept area. Without these between-concept associations, an undesirable gap in lesson development could occur.

Not all the possible discriminations have been identified in this matrix. Others could be identified that would fall closer to the main diagonal listing of rules.

Even first attempts at matrix construction will vastly improve lesson structure and sequence. *The time spent in matrix construction and analysis is the necessary first stage in organizing content.* Though rational analysis of the job/task structure and its supporting knowledges and skills produces an instructional sequence that serves as an acceptable beginning, *there is still the need for empirical verification of the proposed structure.*

Validating Sequence

The criterion test is once again the validation instrument. Giving the validated (and revised) criterion test to a group of about 30 to 50 trained individuals supplies the performance data needed to estimate the validity of the proposed sequence. The trained population can be recent local graduates, or they may come from other sources, as long as the course content is generally about the same. Apprentices who have been on the job no more than six months can also be used in the sample.

Analysis of the test data will reveal the dependence, or lack of dependence, of the successive learning events. The interdependence of the sequential units of instruction, whether between large blocks of curriculum or whether between within-lesson learning events, can be determined by examining two basic situations. *Did the majority of those students who correctly performed a subordinate unit (unit No. 1) also correctly perform the following and supposedly dependent unit (unit No. 2)? And, did the majority of those who correctly performed the higher unit (unit No. 2) also perform the subordinate unit (unit No. 1) correctly?* The same two questions can be asked of the relationship between interim and terminal objectives, between tasks and duties, between concepts and rules, or between the different levels of any series of organized events.

What does it mean when the answers to the questions are in the affirmative? *If 85 percent or more of a sample trained population who perform a certain unit correctly also perform the subordinate unit leading up to it, the sequence is probably correct. And, if 85 percent or more of those who perform the subordinate unit correctly also perform the subsequent unit, the sequence is probably correct.* Note that if only one or the other occurs, there is only a possibility of the sequence being correct. *When both answers are positive, however, there is a reasonable certainty that the sequence is correct.* Of course, there may well be more than one "good" sequence, but under these conditions the designer can, at the very least, be certain the sequencing is not detrimental to learning. There is also the possibility that a reversal of the sequence would not affect performance to any degree and that the two units are equally dependent on a third unit of instruction. By analyzing each in relation to the next unit that comes before the both of them, it is possible to find out whether they both depend on the third subordinate unit.

Suppose the answers to the two questions are negative, what does this mean? *If 50 percent or less (less than chance) of a sample trained population who correctly performed a certain unit failed to perform the preceding subordinate unit, the sequence is probably incorrect*, because performance of the higher unit was apparently not dependent on the ability to perform the preceding unit. *And, if 50 percent or less of those who performed the subordinate unit correctly failed to perform the subsequent unit, the sequence is probably not correct* because performance of the subordinate unit did not lead to performance of the subsequent unit. Note that if only one of the negative situations occurs, there is only a possibility of the sequence being incorrect. But, *when both negative situations occur, there is a reasonable certainty that the sequence is incorrect.*

There is also the possibility of there having been a gap in the training requiring additional instructional steps between the two units if performance of the subordinate unit did not lead to success in the subsequent unit. However, the effectiveness of the

Table 4

Validating Content Sequence

Summary of procedure for analyzing criterion test data from a sample trained population when validating content structure and sequence

Trained Sample (only correct performance)	Performance	Implications
Performs unit (100%)	85% perform sub-unit	Possible correct sequence
Performs sub-unit (100%)	85% perform unit	Possible correct sequence
		Taken together, a certainty the sequence is correct.
Performs unit (100%)	85% perform sub-unit	Possible correct sequence
Performs sub-unit (100%)	50% fail to perform unit	Possible incorrect sequence
		Taken together, indicates bad test item
Performs unit (100%)	50% fail to perform sub-unit	Possible incorrect sequence
Performs sub-unit (100%)	85% perform unit	Possible correct sequence
		Taken together, indicates bad test item.
Performs unit (100%)	50% fail to perform sub-unit	Possible incorrect sequence
Performs sub-unit (100%)	50% fail to perform unit	Possible incorrect sequence
		Taken together, a certainty the sequence is incorrect.

training the sample populations have undergone is not the concern of the designer at this stage. *The main concern now is in determining an effective sequence of learning events.* Later the same sort of analysis will be carried out in judging the effectiveness of the new instructional system.

Sometimes there will be situations where the results are mixed, when one group's data indicate the sequence is probably correct and the other group's data indicate the sequence is probably wrong. What do these results mean? In all probability, *mixed results mean one or more of the criterion test items involved may be bad and in need of revision.*

Table 4, on the preceding page, summarizes the possible situations and gives the implications for each. The process outlined in the table clearly illustrates how the *system content is successively refined through rational analysis of empirical performance data.* The same general process is repeated over and over again in some form throughout the development cycle.

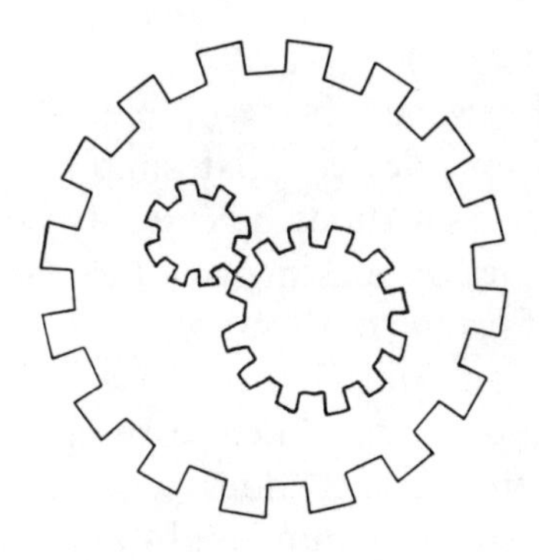

CHAPTER XII

SELECTING INSTRUCTIONAL MEDIA

Introduction

Media selection is an area in which there is little specific information on which to base precise decisions. Although there have been tremendous advances in specialized training equipment and audio-visual devices, *there has been little parallel advancement in the information needed for deciding when, where, and how to use the different media.* Media decisions often are mostly educated guesses, though guidelines can be drawn to help the instructional designer. Although the guidelines are not very definitive, they can at least identify the situations in which the different media are not effective. Moreover, *there is really no feasible way in which the media decisions can be empirically validated against performance data.* The cost in time, manpower, and money usually prohibit the testing and subsequent revision of media alternatives for an instructional system. The test-revise-retest cycle that a system undergoes during production can tell the designer in general terms whether or not the selected media are effective, but not which media might be more effective for a particular situation. As a matter of fact, the final decisions on media are usually based as much on cost analysis as on effectiveness. In a way, that is as it should be—*the least costly medium that can do the job adequately is probably the best choice.*

Media Function

In the general sense of the term, media are merely means of presenting information, devices that simply pass along the information put in them without necessarily affecting either the information or the person making use of them. A student can scan a printed page, half-listen to a lecture, or doze through a training film without learning a thing. *The media merely make information available, they do not teach.* When learning does take place, it is because of what the learner does with the information, not because of any particular medium used to present the information. Of course, certain media can convey certain types of information better than others. If the information is organized and presented in a way that enables the student to act upon and respond to the information, learning will take place regardless of the medium. Learning is primarily a function of organizing and presenting the information so that it will communicate itself to the student; thus, there is the need for constructing an instructional system with two-way communication loops that require the student to respond to and interact with the information. *When system principles are applied, all media (texts, workbooks, films, television, lectures, etc.) become more than just a means of presenting information; instead, the media become a means for eliciting and controlling meaningful learner activity—learning.*

Learning, then, is based on establishing a two-way communication link between the source of information and the individual learner. Oral communication can carry out most of the functions of an instructional system, but it is limited to presenting stimulus situations in verbal terms; thus, objects or pictures may be needed to supplement the verbal descriptions. Oral presentations can be organized into a highly effective two-way communication system (see Appendix K). Printed media can be organized to convey verbal information just as well and often more rapidly than oral communication. Here again, the printed information should be supplemented with pictorial representations of objects, concepts, and principles when verbal description alone proves to be inadequate. *In combination, the different media (oral communication with visuals, printed material with visuals, motion pictures, slide-tape presentations, teaching machines, etc.) gain more capability by taking advantage of the special attributes of each, and thus become more effective than any one medium by itself.*

There also are advantages to be derived through the appropriate longitudinal combination of the different media. The proper phasing of training media, in fact, may be as important as correct selection of such media. If, for example, the performance objective was to operate a certain piece of equipment, an illustrated and annotated description of the equipment could well be used for familiarization and for learning the names, locations, and functions of the parts; this would be followed by a period of training on a part-task procedures trainer, with the training being completed on the operational equipment. The exact period of time to be spent in each of these training phases would be adjusted according to the progress of individual students.

Vocational and technical training places two important limitations on the use of printed material. First of all, the instruction should be conducted under workshop, on-the-job conditions as opposed to a classroom, textbook, and lecture situation; thus complete dependence on printed materials should be avoided. Second, low verbal aptitude and skill on the part of some students can interfere drastically with the learning process. *Particular care must be taken to keep printed materials at the right comprehension level for the students. Frequently, though, a student with a low-level reading comprehension has surprisingly high listening comprehension.* This is one reason why step-by-step slide-tape (teaching machine) presentations and systematized lecture-demonstrations are often more effective than printed material.

There is also a very practical side to developing slide-tape learning units. There are some instructors and curriculum developers who have difficulty expressing themselves naturally in writing. Keeping the language direct, simple, conversational, and in terms that students can grasp is particularly difficult. This problem often can be overcome by recording the conversation as an instructor "talks" a student through a procedure or task. Recording what every instructor does best—helping a student work his way step-by-step through a task—usually provides a highly effective "script" for a learning unit. The script also provides the basis for deciding what (and when) pictures are needed to illustrate each step in the procedure. A series of 35mm slide pictures can then be made to support the taped script. The combined slide-tape presentation in a synchronous projector, which can be stopped and started at will by the student, provides very effective step-by-step instruction. Moreover, that same slide-tape presentation can be converted easily, after trial usage, to the printed booklet format, a less costly medium. However, if a student will later have to rely on complex technical manuals for instructions and guidance when working on the job, the training program must eventually give him the verbal skills he will need to use such manuals. Oral and audio-tape instruction should be used only as a bridge and an aid to the acquisition of the required reading skills. Dependence on oral or taped instruction should be discouraged, because on-the-job situations seldom include such assistance. Also, it should be emphasized that *the combination of printed verbal and pictorial materials, especially when carefully systematized, is often the most effective, most efficient, and least costly medium for most kinds of learning.* Printed step-by-step verbal instructions coordinated with annotated illustrations, and providing for immediate knowledge-of-results to the student as he works with the actual device or equipment, are more than adequate for many situations. A review of some of the performance objectives listed in Appendix A will confirm the generality of this premise.

As pointed out in the earlier discussion in Chapter II of the factors that affect learning, individual differences among learners are considerable and must be compensated for by an instructional system. One view holds that there are qualitative differences among students in their "learning styles"; thus different media are needed by different learners. Another view holds that the same media can be used for all students by permitting each to proceed at his own rate. At present, much more is known about matching media to the various types of learning than is known about matching media to individual "learning styles." Thus, *the best procedure would seem to be to use the media most appropriate to the type of learning involved in each learning event and to provide for self-pacing whenever practical.*

Selecting Media

When instructional system principles are applied to combinations of media, there probably is little difference in the capability of the different media to facilitate learning; a slide-tape presentation, a lecture-demonstration, a programmed text, or a step-by-step demonstration film (if student response and knowledge-of-results, etc., are built in) can all be equally effective for many kinds of learning.

Thus, the choice may not always be between media but between individually paced or group-paced instruction; specifically, how much the pace of the instruction is to be controlled by the individual student. Although self-study situations place the content and pace of the instruction almost entirely under the control of the individual student, group-paced instructional systems adapt to individual needs to a considerable degree. Ideally, of course, the more the instruction is "individualized" the better; however, there are practical considerations that impose limits on how far the designer can go in that direction. The more the instruction is individualized, the more expensive it is in terms of manpower, time, equipment, facilities, and materials, both while the system is being developed and while it is being used. *Completely self-paced instruction is usually more expensive for the very reason that it is more effective than group instruction—each student must have his own material, equipment, facilities, and individual tutoring when he needs it, not when it can be scheduled for administrative convenience, nor to accommodate to the group.*

However, it should be pointed out that the greatest portion of the higher cost is associated with the preparation, the production, and the start-up of an individualized course. Initial costs will run higher than those of a comparable standard course at the outset, but long-term costs (amortized) will be no higher, and stand a good chance of being lower. The amortization rate will be higher because more and better trained graduates can be turned out in less time (typically about one third less). Thus, *despite initial high costs, individualization can produce long-term financial gains as well as long-term educational pay-offs.* Ingenuity in applying the systems principles to group instruction can produce learning situations that approach the effectiveness of self-paced instruction. As the number of students involved in simultaneous instruction goes up, the costs per student go down proportionately. *Thus, the system designer can balance cost versus effectiveness for each unit of instruction.* Appendix K discusses in some detail the advantages of a systematized group-paced lecture-demonstration and gives a sample script for that mode of presentation. In addition to the obvious cost implications, the script or lesson plan for a group-paced instructional system is often the logical first step toward individualization. Systematized group lecture-demonstrations (as described in Appendix K), in combination with individually paced and controlled self-testing and individually administered criterion performance testing, constitute a giant step toward the total instructional system. Once the course material has been developed to that degree, the final step to complete individualization is relatively easy. Note that a single mode of instruction or medium is not decided upon for the entire course, nor even for an entire lesson, but different mixes of media and mode are chosen to fit the particular objectives within each unit of instruction. *The training objectives, by specifying the performance capability to be learned, and indirectly the type of learning involved, also specify the most appropriate media mix.*

Training Aids and Training Objectives

The different kinds of learned performance were described in detail in Chapters III and IV, and the close relationship of training objectives to the types of learning was discussed in Chapter VIII. *The information in this section is organized to help the system designer select the training aids most suited to specific training objectives.* Each category of training objective (based on type of learned performance involved) is listed and the most effective training aids are then discussed in summary form. Note that the aids discussed under each type of training objective are listed in order of decreasing effectiveness and suitability.

Learning Identifications and Locations

Transparencies. Transparencies (slides, overheads, etc.) are very effective primarily because of the extensive flexibility in the manner in which items of equipment may be depicted. Cost and preparation time factors also are quite favorable for transparencies.

Training Charts. As an aid to meeting this training objective, training charts differ very little from transparencies. For presentations at a number of training locations, training charts may be preferred over transparencies because there is no need for projection equipment.

Simulators, Procedures Trainers, Mock-Ups, Teaching Machines. Any one of these aids may be used for teaching identifications and locations. However, because of cost considerations, they generally cannot be justified in terms of this training objective only.

Learning Skilled Perceptual-Motor Acts
(motor chaining)

Simulators. Simulators are quite useful for this training objective if the control-display relationships are presented with considerable fidelity. This is particularly true where the action being learned is a continuous-control tracking activity.

Procedures Trainers, Part-Task Trainers. Procedures trainers are appropriate for perceptual-motor acts which are weighted more heavily toward procedural components than toward continuous-control components.

Learning Procedural Sequences
(verbal and motor chaining)

Procedures Trainers. Procedures trainers and part-task trainers, as the names imply, are training aids designed specifically for this training objective and consequently should be quite effective.

Simulators. Simulators can be as effective for this training objective as procedures trainers. However, considerations of cost and economy of student and instructor time frequently dictate the use of procedures trainers for learning sub-task procedural performances.

Training Films. Training films can be quite effective in teaching procedural sequences if the student is given the opportunity for step-by-step practice on a procedural trainer of the desired responses as they are being presented on the screen.

Learning Discriminations

Training Films. Training films can be used in teaching perceptual discriminations when most of the required cues are visual. For the most effective use, the training film should present pictures of the actual equipment operating within a realistic environment.

Simulators. Simulators are appropriate for teaching perceptual discriminations if all requisite cues can be presented within the simulator. Be sure to include all the important cues, however, because some of the cues underlying certain discriminations may, on the surface, appear unimportant.

Transparencies. Transparencies are appropriate for this training objective if they are realistic photographs of a situation and include the major identifiable visual cues underlying such discriminations.

Learning Concepts, Principles, and Relationships

Television, Simulators, Animated Panels, Training Films, Operating Mock-Ups. All training aids and devices which have the capability of presenting or illustrating motion characteristics are quite useful for the teaching of functional relationships among concepts, principles and operating parts of machinery, etc. This is the training objective most frequently encountered in vocational and technical training.

Transparencies, Charts, Teaching Machines, Procedures Trainers, Non-Operating Mock-Ups. All of these training aids can be used to illustrate principles and functional relationships of equipment operation if structured in terms of this objective.

Problem-Solving and Trouble Shooting

Teaching Machines. Teaching machines appear most appropriate for this objective when they are programmed so that the problem is presented in conjunction with all required information items underlying proper decisions. The student then may practice arriving at appropriate decisions. For each decision selected, the student may receive immediate feedback as to the adequacy of his decision and reasons why it might not have been the preferred decision.

Simulators. Simulators, particularly those classed as full system simulators that incorporate complete system capability, are appropriate for training in decision-making and trouble shooting responses.

Actual Equipment. Actual operating equipment such as radios or engines are often not as effective as simulators because of the difficulty in inducing and controlling realistic malfunctions.

Training Films. Films can be effective for this objective, but only if they present problem areas requiring decisions within a realistic operational context.

Almost without exception every unit of instruction will need the support of training aids and devices of some sort. Training equipment has considerable influence on both the effectiveness and the cost of a training program; so it is important to make sure the selected training aids and devices are fully justified. *Remember, the simplest medium, or mix of media, that can adequately support the training objective is the most practical and probably the best choice.* For example, a short single-concept motion picture sequence augmented by a series of still pictures and text can often be more effective than a lengthy training film. Or, a series of stop-motion still pictures can be used to portray continuous motion just as well as an actual motion picture segment. And, in most instances, printed step-by-step directions with sequential illustrations can be as effective as a slide-tape presentation for learning complex maintenance procedures and equipment operation.

General Guidelines

The instructional system designer faces many intricate decisions when selecting media and, because there are so many factors, both known and unknown, the decisions are often discretionary and judgmental. The preceding discussion has attempted to provide some helpful insights and suggested solutions to the problem, but a summary of general guidelines will be useful at this point:

- The training objectives themselves are the primary determiners of the best media to use.
- Matching the media to the various kinds of learning involved in a lesson is more effective than trying to satisfy the so-called individual *learning styles* of students.
- There usually is no single best medium, or mix of media, for a unit of instruction, because each learning event leading to the attainment of the objective may require a different mix of media.
- The best procedure is usually to select the least elaborate and least costly medium that apparently will enable the learner to acquire the desired capability.
- All things being equal, well illustrated, step-by-step verbal instructions with feedback to the student are the most practical, effective, and efficient medium for most types of learning.
- In some instances, cost factors may dictate media selections that are concerned primarily with arriving at the most advantageous mix of self-paced individualized instruction and group-paced lecture demonstrations.

The procedures and guidelines put forth in the preceding

discussion, recognizing their inherent limitations, *should enable the instructional system designer to choose predictably effective media for each learning event leading to the attainment of a training objective.*

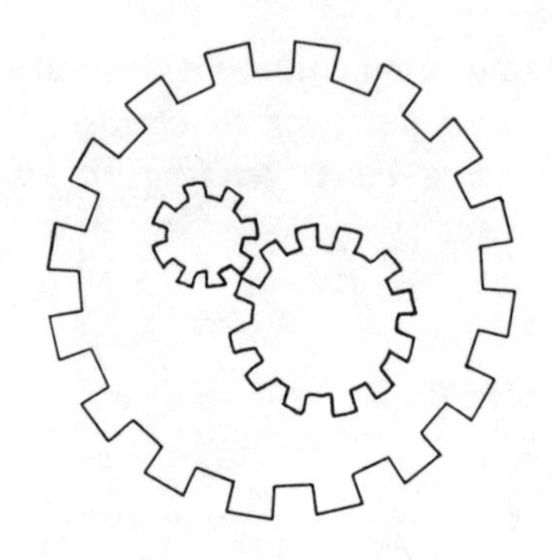

CHAPTER XIII

PRODUCING AND VALIDATING THE SYSTEM

Introduction

This phase is the culmination of the development process. Everything now begins to fall into place almost of its own accord. All the detailed specifications have been established, all the major design decisions have been made, and all blue-prints have been drawn up—everything is in readiness for producing a test model of the system. Before production can start, however, an important document has to be assembled and submitted to management for approval.

The System Development Plan

This document is the first of two quality control instruments produced during the instructional systems development process, the second being the Instructor's Manual. Actually, *the System Development Plan is a collection of all the quality control data of any importance to training managers* that is available at this point. The contents are as follows:

1. The completed task analysis summary forms.
2. A list of the validated training objectives in validated

sequence, supported by a summary of the validation data.

3. A list of the validated criterion test items in validated sequence, supported by a summary of the validation data.
4. An outline of instructional strategies with the associated content (objectives) identified.
5. The production and testing plans for the system, including time schedule, resources assigned to do the job, source and number of students to be tested, etc.

The information contained in these documents is direct proof (or disproof) of the adequacy of system design. The System Development Plan should be made available to training management as soon as possible, because the decision to proceed or not with the production and testing of the model system is based on their evaluation of the information in this document. (This is the second go, no-go decision point for training managers and administrators, the first being their evaluation of the Training Feasibility Study.) *Once the plans have been approved and the decision has been made to proceed with production, the System Development Plan becomes the principal source of information for the production team.*

Design and Format of Learning Units

Although the outward format of learning units and lessons may vary considerably, the internal design should always include the following structural elements:

1. An exact description of what the student is to do as a direct result of the learning activity called for by the lesson (the performance objectives).

2. A statement of the function and applicability of the knowledge and skills to be gained from the unit.

3. A list of tools, supplies, equipment, training aids, technical and service manuals, textbooks, etc., that are needed by the student to carry out the prescribed activities.

4. A step-by-step, self-paced learning activity guide for the acquisition of the skills and knowledge specified by the performance objectives.

5. A means of providing interim progress checks and self-evaluation with immediate feedback for the student.

6. An instrument, capable of serving as a pre-test and/or a post-test, with which the instructor can evaluate, certify, and record the attainment of the terminal performance objectives.

The overall length of each learning unit is, of course, determined by the objectives of that unit. However, no learning unit should require less than a full class period to complete. If the majority of the students can complete the unit in less than a period, the objectives are probably too trivial to be truly terminal objectives and are actually only sub-objectives. Most units will probably require several periods to complete (typically a two- or three-hour shop period). The relatively long length of some of the learning units should not affect student motivation adversely, as long as the activities are organized around visible and attainable sub-units and sub-objectives.

As emphasized earlier, behaviorally stated performance objectives and their derivative criterion test items specify the content, method, standards, and setting of the learning activities. Thus, the learning units, of necessity, must create a situation that calls for a realistic, hands-on, job performance activity—both for learning and assessment purposes. The learning materials must prescribe job-like learning tasks whenever possible, relying heavily on workshop activities. Supporting knowledge structures must, of course, be included in the learning units, *but the acquisition of concepts and principles should be integrated with development of the hands-on skills—not treated separately from, nor prior to, the job skills they support.*

Within the confines of the above guidelines, the curriculum developer is still left with considerable latitude as to content and format of the learning units. The choices range between developing completely new, highly structured materials that are, in effect, programmed instruction packages, and writing brief learning activity guides that require the student to make extensive use of off-the-shelf materials such as technical handbooks, service manuals, and available textbooks. Very practical reasons such as the availability of time, money, and skilled curriculum writers often determine the format of the learning units. Obviously, *the more the learning units make use of off-the-shelf materials, the less time, money, and effort are required for development.*

Just as in the case of media selection, there is no single learning unit format that is best for all objectives. Format should vary to conform with the requirements of the performance objectives. For instance, if the objectives and the on-the-job situation call for the actual use of service manuals, technical handbooks, or operating instructions, the learning unit itself should require the student to learn to use them in carrying out the job task. Thus, even when completely new material is being written, some of the learning units may call for student use of off-the-shelf technical materials.

As pointed out earlier, instructional systems development is always a series of compromises in which trade-offs are continuously being made. If the development process has been carefully and conscientiously carried out up to the point of learning unit development, there is room for considerable compromise at this stage. Adjustments to meet the constraints of time, money, and manpower can be made without seriously affecting the product. However, all the structural elements of good learning unit design must be included (the six elements outlined at the beginning of this section) if the product is to be an effective vehicle for learning. Of all the elements, *probably the single most important component is the inclusion of valid and reliable performance evaluation devices which permit the student to check his own progress intermittently and to demonstrate his final achievement of the objectives.* Full use must be made of criterion testing procedures within each learning unit, because the built-in evaluation devices, in fact, provide the structure and organization of the learning activities.

When practical and appropriate, the performance evaluation devices should serve the following purposes:

1. An objective test of the supporting knowledge structure for the performance tasks.

2. A check list of critical criterion tasks and sub-tasks that must be performed to achieve the objectives. (See Appendix F.)

3. Both the knowledge test and the criterion check list must be self-scoring and must provide immediate knowledge-of-results.

4. The knowledge test and the criterion check list should be constructed to serve as a diagnostic pre-test/post-test and learning activity guide. Performance on the tests determines what the student needs to do to achieve mastery of the objectives. (It may be that he can skip the unit entirely.) Where there are indications that a student may be capable of by-passing a lesson, he should be given the objective knowledge test first to see if he is ready to go directly to the performance test. Unless he can show reasonable mastery of the knowledge structure, he probably would be incapable of successfully completing the more critical and time-consuming performance test.

5. Both the knowledge test and the criterion check list should serve the student as intermittent self-checks on his progress in preparation for instructor certification.

6. The combination of the knowledge test and criterion task check list should provide the instructor with an objective means of determining whether or not the student has achieved the performance objectives.

7. The criterion task check list also serves as the basis for drawing up an "Occupational Readiness Record" which certifies and specifies the student's capabilities to potential employers.

Regardless of the format, content, and strategies of the learning units, the performance evaluation devices remain the key to their effectiveness.

A series of sample learning units is provided in the Appendices (Appendices H through K). The samples range from a brief and minimal learning activity guide, that relies heavily on off-the-shelf materials and adjunctive performance evaluation, to the more sophisticated, self-contained and highly structured "programmed" learning units.

Production of Instructional Units

The instructional system is not produced in its entirety and then tested; rather, it is tested as it is being produced. The system is produced and tested unit-by-unit until it is completed. Lessons or instructional units of one to three hours in length are produced and tested separately. The testing is not a one-time event either; instead, each unit is tested first on only two or three students, then on six to ten, and finally on at least 30 to 50 students, and between each testing the unit is revised to bring it up to standard. *Essentially, the system production phase is a continuing cycle of produce, test, revise, retest, etc., until the criterion test data indicate that the prescribed standards have been met.*

During the production phase, all the system training materials are developed—texts, study guides, workbooks, and the supporting training media. Because the instructional materials must communicate well enough to be understood by the majority of the students, the training system must first of all go through an editing process. The materials must proceed through three kinds of edit, each of equal importance. First, there has to be an edit for technical accuracy by subject matter experts. Regardless of the content, the form of presentation, or the style of language, the material first of all has to be accurate. *As is true with the other kinds of edits, technical editing is a continuing process; that is, the experts must continue to check the technical accuracy each time the material is revised during the development cycle.*

The second edit is for composition. This is true whether the material is on tape, in a text, in a script, or whatever kind of verbal communication is used in the system. Does the material appear to be capable of communicating with the intended student population? Does it use the language of the future learner, or is it written for those already expert in the subject? Is it simple, straight-forward, conversational in tone, and free of excess verbiage, unnecessarily long words, and ambiguity? Are the explanations clear and concise? This review is much more thorough than looking for and eliminating split infinitives, dangling participles, and misplaced modifiers. *The composition edit is an attempt to assure that the material will communicate with its expected audience.* The students themselves will, of course, be the final judges of whether the materials communicate when they try out the materials.

The third edit is concerned with the application of the elements of an instructional system. It is not enough for the instructional materials to be technically accurate and to be clearly expressed; the materials must make full use of those techniques that will create conditions that will be conducive to learning. *The primary purpose of this edit is to attempt to prejudge how well the instructional units are patterned after the detailed blue-print set forth in the System Development Plan.* Does the material appear to be following the stated objectives? Does the inter-unit and intra-unit structure follow the sequence dictated by the Course Development Plan? Does the material appear to take full advantage of the preplanned strategies for learning? Regardless of the kinds of training material being produced, the designer must build in the attributes of an instructional system as described in Chapter V. Whether the selected medium for a unit of instruction is a lecture-demonstration, a filmstrip, a motion picture, or a combination of some or all of these, the designers must somehow apply the elements of an instructional system. The designers must always strive toward the ideal of the adaptive, closed-loop communication system. *Of course, the efficiency and the effectiveness of the instructional material cannot be judged finally until the materials are actually tried out on sample students.*

Individual Tryout

Each unit of instruction is first tried out on an individual basis with only two or three sample students who are representative of the target population. The designer should pick only those individuals who fall within the range of aptitudes, prior knowledge, skills, background, and attitudes displayed by the typical student. If the sample does not fall within this range, the results will be biased and any generalizations from the sample to the target population will be in error. However, *the sample students used in the first tryouts should not be average; in fact they should come from the upper 25 percent in aptitude and background.*

There are several reasons for choosing upper-level students. The brighter students often can help point out and analyze weak spots in the instruction. Moreover, if the better students cannot learn from the material, the less capable student certainly cannot either. On the other hand, if low-level students are tested first, and they do well, there is no way to tell if the material is too easy, if there are too many steps, or too much practice. It is more economical of time and effort to work down from a known point of difficulty than to work up from an unknown point of difficulty. Moreover, it is easier to add steps to lower the difficulty level than it is to delete material to make it more difficult.

During the individual tryouts, each student works through the unit of instruction in the presence of the designer. *The student should understand that the observer is not evaluating him; but that he, the student, is evaluating the instructional material.* It is very important that the student understand his role in the evaluation. The designer should encourage him to comment freely about the instruction; especially at the places where he has difficulty. The designer should not help the student during the tryout unless it is absolutely necessary to progress. In any event, the observer should keep careful notes so he can locate the problem areas later. When the student does encounter difficulty in understanding or fails to respond correctly, the designer should try to get his reactions and thinking about why he had trouble at that point. Asking the student for his suggestions on how to improve the instruction can be helpful too. When the student has completed the unit, he is given the criterion test for that unit to provide concrete data on his performance.

Although the student's comments and his performance on the criterion test may indicate some real weaknesses in the unit, the designer does not, at this time, revise the content except to correct technical inaccuracies or other very obvious deficiencies. *Normally, at least two or three individual tryouts are conducted before significant revisions are made.* Consistent trouble spots and errors on the part of several students are what the designer should be looking for. If the others in the subsequent tryouts have similar difficulties at the same places, it is a strong indication that revision is needed.

An even better guide as to where and how the material needs revision is the performance data from the criterion test. Clusters of errors definitely call for revision. Analysis of the error patterns helps to pinpoint the problem areas. *If the students fail to perform a particular terminal objective on the criterion test, the instruction leading to that objective needs revision, and failure to perform some of the interim objectives leading to that terminal objective indicates exactly where the breakdown occurred.* If the need for revision is apparent after two or three individual tryouts, the designer must then make the necessary changes before proceeding with any more testing.

Small-Group Tryout

If the instructional unit under test proves to be in need of a great deal of revision, it is probably best to try it out again on two or three more students on an individual basis before proceeding with group testing. On the other hand, *if the unit is fairly effective and does not require too much revision, the unit should then be given to a group of about six to ten students who cover the range of abilities of the target population*—a few each of the top, middle, and lower ability ranges. The instruction should be given to the group under conditions that approximate the actual training situation. Criterion test data are again analyzed to diagnose and locate trouble spots and the content is again revised. By now student performance on the criterion test should be approaching the desired level—each item on the test should be performed correctly by at least 85 percent of the students. The designers should be reasonably sure now of the effectiveness of the unit; so the final developmental tryout of the unit on a large group can be undertaken next.

Large-Group Tryout

The final step in the validation process is to try out the unit on a group of 30 to 50 students under conditions that are as close to the actual training situation as possible. The large-group tryout is conducted by the development team, but the regular instructional staff should assist as required. Being present and assisting during the group tryout will give the instructors a chance to start becoming familiar with the system they will soon be operating. Again, the students should clearly understand that they are helping to evaluate the system; that they are not being tested even though they will be taking the criterion test.

There are several reasons for this final large-group tryout. Even though the unit has already demonstrated its effectiveness and has been revised based on the performance sample students, *data from a group of 30 to 50 are needed to verify the preliminary findings.* The designers can have considerable confidence in the results of any analysis of the data from a sample population of this size. Thus, the data will provide a solid base for final revision and refinement of the instructional unit. Giving the unit to a large group will also give the designers a chance to work out any administrative, equipment, or facilities problems that might cause trouble later when the system is turned over to the instructional staff for implementing. One important administrative consideration that must be worked out is the amount of time that has to be allowed for the unit in the total curriculum. *A reasonable time allowance for a unit of instruction is the average time taken by the first 85 percent of the sample population to complete the unit.*

The goal of the designers is to produce instructional units which are all but guaranteed to enable 85 percent of the students to achieve all the objectives; specifically, 85 percent have to get 100 percent on the criterion test. Actually, the goal is not at all unrealistic for an instructional system. Remember, within reason, there are no rigid limits placed on the length of time the students can take in achieving the objectives. Also, many of the training objectives set reasonable and realistic standards of performance that allow for some error. The chances for success with the large-group tryout are high because empirical validation has been required at each of the prior development stages. Regardless of the outcome of the large-group test, however, a complete analysis of the performance data should be carried out. The same analytical procedures prescribed at the end of the chapter on "Organizing System Content" (see Table 4, Chapter XI) should be completed as a final check on contents, sequence, and method. *Even though the unit tests out at the 85 percent level, or even if it does better, the unit is revised once again to repair any weakness that shows up in the analysis of the data.* Although highly unlikely, if major revision is needed at this stage, there must have been a breakdown somewhere during the earlier validation procedures.

With the completion of this final test-revise cycle for each unit in the system, the development process has come to an end. The role of the systems designer has not ended, though, for he has an important part yet to play in implementing and field testing the system.

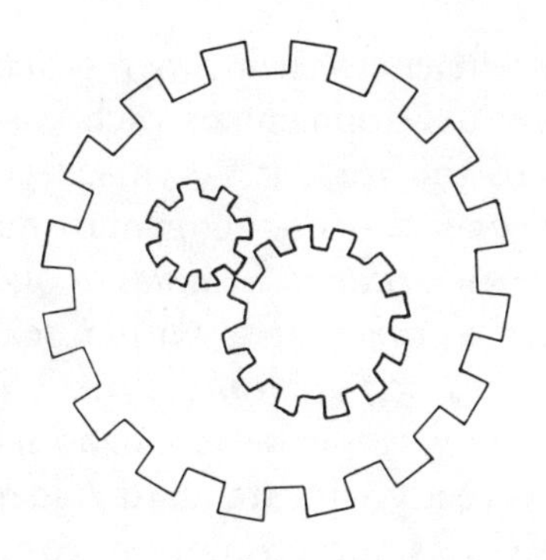

CHAPTER XIV

IMPLEMENTING AND FIELD TESTING THE SYSTEM

Introduction

Many well-designed instructional systems have not lived up to the expectations of the designers because of the lack of specific plans and procedures for installation and implementation. There can be no one set guideline for implementing the wide variety of vocational training programs in all the different schools; only general guidance can be given with such diversity.

It is not the intent of this chapter to supersede the guidance and the various policy instructions issued by state departments of education, the local school districts, and other agencies. Rather, this chapter will discuss the instructor's role in an instructional system, some guidelines for developing the Instructor's Manual for each course, and some potential problems in system operation. In addition, the policy and procedure for field testing a training system will be outlined at the end of the chapter.

The Role of the Instructor

The instructor must recognize that his primary goal is to provide for the continuous intellectual, occupational, social, and emotional development of all students, without interposing artificial or arbitrary barriers to their progress. The instructor must promote the optimum development of each student in each area, but only in relation to the student's own self-perceived potential. To succeed in these goals, the instructor must maintain a situation in which *the individual student assumes a major share of the responsibility for his learning* and, within reasonable limits, is allowed to proceed at *a self-determined rate, using self-selected modes of learning to achieve self-selected and self-satisfying goals.* These ideas contrast directly with standard graded school organization and such practices as non-promotion, common achievement standards, graded curriculum, grading "on the curve," large-group instructional methods, graded textbooks, and instructional media and materials that provide common content to be learned at the same time by all students.

The instructor is a vital part of any instructional system; in fact, he remains the key to successful learning. Although an instructional system may typically relieve the instructor of many of his former tasks, and may change his role considerably, *he is, more than ever, the man who makes the system "work."* He must be sure that the objectives of the training program are clear to the student, he must motivate them to actively participate in the training situations, he must guide student activity throughout the course, and he must continually assess and analyze both student and system performance. Under the systems concept, *the instructor truly becomes a manager and facilitator of learning.*

In an instructional system, the instructor has to serve in two essential roles: as course administrator and as an individual tutor and counselor. As course administrator, the instructor must be sure that the required training supplies, tools, and equipment are available. He must monitor the activity of all students, making sure they have completed the required work and that they have done it correctly. He must administer the various evaluation devices to determine the student achievement of course objectives, and he must gather and tabulate all the performance data of both the student and the system. The instructor must keep track of student progress, take attendance, help with extracurricular activities, etc.—all the myriad details involved in making sure the system keeps operating. *As important as the administrative tasks are, the instructor's role as personal tutor and counselor is certainly more important;* and not incidental to that role is providing motivational reinforcement and encouragement where needed. The instructor's role as a personal tutor may be new to many classroom instructors. A combination of re-education concerning the instructional systems concept and retraining for this new role as a part of a system may be needed for some instructors.

The main goal of an instructional system is to have at least 85 percent of the students achieve 100 percent of the objectives. It is, of course, almost impossible from a practical standpoint to develop a training system that will anticipate and provide for all the needs of all the students, all of the time. Instructional systems designers should recognize this fact and not try to achieve perfection. *Practical systems can be designed to take care of a majority of the individual differences, but the task of providing for all the differences among the students can best be left to the instructor.* Some students will always have trouble meeting certain objectives, understanding certain concepts, and performing certain tasks. Trouble spots will differ among the students. The instructor must be alert for these trouble spots and be prepared to give individual assistance where needed. In addition, the instructor will have to deal with the problems of the students who are overly fast or overly slow, for no system can take care of the extremes in ability. The instructor must be careful not to do the work nor the thinking for the student; rather, he should individualize the instruction to meet each student's needs. Thus, he must first diagnose the student's problems based on observation of student performance and careful analysis of criterion test data and then he must provide appropriate assistance. *Because an instructional system takes care of most of the needs of most of the students most of the time, the instructor will have the time to help the 10 or 15 percent of the students with special problems.*

At first, there is a tendency for some instructors to feel uncomfortable in their new role of personal tutor and counselor. In some cases they may even revert to being classroom "lecturers" only. For example, when one or two students have a problem, some instructors will stop the whole class, step to the blackboard, and start lecturing—performing the one role they know best. *The instructor should not help one or two students at the expense of the others in the group*; instead, he should provide individual help for those who need it. Finally, the instructor should make note of the possible weakness in the system and later analyze that portion of the lesson to see if it should be revised to prevent further problems. The list below summarizes the many tasks the instructor must carry out in his new role as the manager of an instructional system.

- Diagnose individual learning needs
- Prescribe appropriate learning experiences
- Provide appropriate learning materials and equipment at the appropriate time
- Test and evaluate individual progress and achievement
- Compile individual and group progress and achievement records
- Provide individual tutorial and counseling help when needed
- Provide motivational reinforcement
- Provide supplemental or alternative materials and experiences.
- Assign, schedule, and coordinate individual, small-group, and large-group learning activities
- Schedule, coordinate, and control the assignment of learning materials and equipment

- Evaluate feedback data on the effectiveness of learning materials, equipment, and methods

Central to the aim of individualized instruction is *the promotion and development of self-directed learning, because true education is a continuous process of expanding the individual's capacity and desire for further learning.* Learning does not occur simply because an instructor discovers and nurtures a student's inherent abilities. Instead, learning results from his interactions with the instructor, other students, and learning experiences—all contributing to the growth of both his capacity and desire for further learning.

Because the generalizable skills of decision-making and problem-solving are integral to the capacity to learn, the instructor should encourage each student to participate fully in the decisions concerned with his learning program and progress. Each student, in cooperation with the instructor, should be allowed to make decisions concerning what to learn, how to learn, and when to learn. Within reasonable limits, each student should choose for himself the goals, the mode, and the rate of learning that are commensurate with his self-perceived capabilities. Individualized learning occurs only to the extent that each student, beyond certain minimums, is allowed to progress at his own pace; to choose the learning materials and mode suited to his own needs; and to vary the content and sequence of learning objectives to meet his own goals. *Providing the student with the opportunity to assume real responsibility for his learning, and thus to develop self-management skills that will be needed throughout his life, is perhaps the single most important contribution the instructor can make.* (See Appendix G, discussion of self-testing.)

Not only must the instructor help the students to develop a sense of responsibility for their own learning, but he should also encourage them to assume the responsibility to help each other in the learning process. Each student should be allowed to act as a tutor in the areas in which he feels competent. There is considerable evidence to show that students can often learn best from their peers. Thus, students of all levels of capability and at all stages of progress should be allowed to work together in the classrooms, the laboratories, and the shops. Serving as a tutor forces a student to face the reality of his own degree of capability, especially his ability to explain and to demonstrate, which is the real test of competency. Both the student-tutor and his "pupil" obtain benefits from the process. Therefore, the instructor should actively encourage the students to seek help from and to give help to each other. (See Appendix N, "Class Management System.")

Course developers must consider the needs of the instructor and his function as a key part of the overall system. The best way to insure effective instructors is to provide them with the necessary information and training for their roles as administrator, tutor, and counselor. *Specialized instructor materials and training should be included as part of course design and development, not as an afterthought.* Course developers should provide Instructor's Manuals and Master Lesson Plans as an integral part of the instructional system material.

Preparing the Instructor's Manual

A great deal that has gone into the development of a training system does not show on the surface. In particular, the capabilities and limitations of a system are not readily apparent to the instructors who will be a part of the system. This information is critical for implementing the system and must be furnished in an Instructor's Manual.

The information presented in the manual must show the school staff how the system was developed, what specifically it teaches, how effective and efficient it is, and how it can best be implemented. The topics that should be covered are as follows:

Course Description

The following information should be included in the course description:

1. The course title and a brief statement as to the purpose and scope of the course—for what jobs it prepares the student and where it will lead him.

2. A concise outline or brief overview of the contents of the system.

3. A physical description of the training system—number and length of lessons, student and instructor materials, training aids, equipment, and tools required.

4. Appropriate comments concerning the training methods and techniques used in the system.

Population Description

The population description should describe the student for whom the system has been designed and developed. It should contain the following prerequisite information:

1. Age and educational level, including reading and mathematics, of the student population.

2. Previous vocational training and related knowledge.

3. Physical and personal characteristics required by job.

Performance Objectives

This very important part of the manual should furnish a list of all the behavioral training objectives for the course, both interim and terminal.

Criterion Tests

Copies of all the criterion tests should also be included in the manual. Answers to the tests should appear either on the tests themselves or on separate answer sheets. Criterion test items should be cross-referenced to the behavioral objectives. It should be clear to the instructor just which objectives are being tested by which test items. It is particularly important to include detailed descriptions for each of the "show and tell" performance tests the instructor will have to conduct.

System Performance Data

This part of the manual should contain a thorough discussion of the steps taken to test the effectiveness of the system and the data resulting from individual and group testing. Also included should be a description of the students used to validate the system—number of students, method of selection, age, and background data. An account should be given of the conditions under which the tests were given. Appropriate comment should be made concerning the adequacy of the training system in meeting industry training standards. Tables should be included to show comparative pre- and post-test results and other performance data. Every effort should be made to describe completely the results of the validation process.

Administering the System

This portion of the manual should describe how the system can be most effectively and efficiently used. No single list can satisfy the complete range of training programs, but the following items are suggested.

1. Motivation information and techniques relating to the specific system and future job possibilities.

2. Materials required by the students, including tools and equipment.

3. Guidance of student activities and the role of the instructor.

4. Recommendations for associated activities. These should include methods for making the training more realistic to the student. Also included should be suggestions for incorporating and integrating other school activities into a total training experience for the student—relating this training to basic education and work projects.

5. Recommendations for handling individual student differences—the special activities and materials that should be provided for fast and slow students.

6. Instructions for test administration to insure comparable graduates from all classes.

7. Special instructions for training the instructors who will actually conduct the course.

Implementing the System

Many feel that, once developed, an instructional system is a finished product and that no problems should arise during its installation. This is far from true. Problems will occur that are peculiar to individualized instructional systems, and some problems will arise that would come up in any school situation.

Where portions of the total training system involve self-paced instruction, course managers must be alert to possible scheduling problems. If some students finish part of the instruction early, *plans have to be made to accommodate those who finish early.* Students can be motivated to complete the course quickly by giving the faster students special privileges, tutoring responsibilities, relief from attending class, taking a part-time job for credit and pay, or any of a variety of tangible and intangible status rewards to encourage them to work up to their capabilities.

One word of caution, however; unless special care is taken, it is also possible to undercut the advantages of self-pacing through improper administration. If the student who completes the course or a portion of the course early is assigned extra work for the lack of something better for him to do, if he is made to take part in "busy work," or merely given more of the same to do while waiting for the other students, then there is no motivation to move ahead more rapidly than the others. Another problem is the occasional student who races through the training material, paying less attention to the quality of his work than to his rate of progress. The instructor must be sure he is actually achieving the objectives (criterion performance tests) rather than just superficially rushing through the material.

Developing an appropriate set of administrative procedures for dealing with students in an individualized or self-paced training program requires a great deal of thought and planning. Considerable experience has been acquired at some schools in administering individualized training programs in basic education. Mathematics and reading programs are two good examples of the potential of self-paced instruction. Some sort of "pipeline" scheduling system, in which the new enrollees are fed a few at a time into training as needed to fill up the space left by someone finishing at the other end of the "pipeline," seems to work best. Cooperative work-study programs offer an ideal solution to both the administrative and the motivational problems associated with the "early-finishers." Rewarding the students who finish the study portions quickly with increased participation in the work programs takes care of the early finishers both administratively and motivationally. In any event, *scheduling procedures must remain highly flexible to meet the needs of a wide range of ability and the continually shifting individual rates of progress.*

Another problem is a changing student population. Continuous checks must be made to catch any overall changes in entry knowledge and skill, particularly in reading and mathematics. An instructional system is designed to train a particular kind of student; and if the qualifications of the majority of the entering students change radically, the system will have to be modified to meet the change. *To be truly an instructional system, the content has to be revised and validated continuously to adapt to any changing needs of the students.*

Another potential problem in any training program is student attitude. Unless the operation and the goals of the training system are fully understood and appreciated by the student before the start of the course, attitude problems may arise that could otherwise be avoided. Just as important is instructor attitude. Many recent developments in educational technology apparently ignore the instructor or seemingly minimize his role. It is important to stress that the instructor is the key to success of training systems. *Any instructor who does not understand the importance of the role he plays in an instructional system, and acts accordingly, can seriously weaken its effectiveness.* Managers and course supervisors must be certain that each instructor introduced into the system is given clear guidance for his role. Course developers must insure that instructor materials are the best possible. All too many early systematizing efforts have overlooked the need for instructor guidance, for instructor materials, and detailed lesson plans, especially for those lessons that are to be group lecture-demonstrations. In many cases, fully developed student learning materials can serve as the lesson plan for the instructor as well. Lesson plans may be included in the Instructor's Manual, but it is probably best to provide them separately.

Preparing Master Lesson Plans

Master lesson plans serve several purposes. They serve as a check list to help the instructor with the final development of each group lecture-demonstration. They can be used as a teaching guide during the lesson. Master lesson plans serve as a common reference point for all instructors who teach the same topic. *The master lesson plans serve as a step-by-step guide for the instructor for planning and conducting his group demonstrations.*

The first part of each lesson plan should be an overview of the topic, the individual performance objectives, references, tools and equipment lists, and related general information. The second part should contain a complete outline of the lesson development. No particular format will be prescribed here. (See Appendix K for a sample Group Lecture-Demonstration lesson plan.) However, the following suggestions are made for preparing master lesson plans.

- Indicate exactly how each training objective is to be carried out, specifying methods, materials, and media.
- Include an estimate of the amount of practice and the amount of time required for each area of student activity.
- Make a special effort to identify exactly what the

instructor will be doing during each phase of the lesson.

- Provide detailed instructions for the conduct of the student practice sessions.

- Indicate when and where student practice materials and equipment are required.

- Make sure that lesson activities are student-centered, spelling out both the instructor's and student's activities.

- Plan the reinforcement schedule.

- Give special instructions on the use of the training aids.

- Provide for effective evaluation of student achievement of the objectives.

- Make definite suggestions on when and where to review and evaluate student activities.

- Provide for an introduction and summary for each lesson.

Good master lesson plans tell the instructor what he has to achieve during the lesson, suggest how he can best reach the goals, but leave many of the details of how to get there up to him. The instructor, as part of the system, must be encouraged to use his own imagination and initiative to constantly better the performance of the system. *No training system will work unless the instructors "on the firing line" are deeply and totally involved in the process.*

Field Testing

Field testing, as the term implies, is the evaluation of the training system under "field" or classroom conditions. Up to this point, the evaluation has been part of the development process, and the testing has been carried out under formal, almost research-like, conditions. Furthermore, the testing and evaluation has been under the direct control of the system designers and conducted under nearly laboratory conditions. Thus, *the field test is really the final phase of the systems development process.* Although there is no formalized testing in the usual sense, this final evaluation is crucial because it is the first time the system has to stand on its own. The operation and evaluation of the system is left in the hands of the users—the instructional staff and the students. While the instructional staff may have participated in the development process to some degree, they have been mainly assisting the course designers. Now the roles have to be reversed: the instructional staff has to be in full command, with the course designers assisting them. No matter how well the system performed during development, it has to satisfy the users in the final analysis.

There is no formal field trial as such; in fact, there is really no final test of the system. Rather, the system is monitored, evaluated, and subsequently revised on a continuing basis as long as it is in use. Careful analysis of the performance data derived from the criterion tests is especially important during the early stages of system installation. Staff and student comment should also be taken into account at this stage. However, at least the first class or the first 30 or so individual students should complete the course to allow for staff familiarization and system shakedown before any actual evaluation takes place.

Once the shakedown period is over, performance data on at least 30 students should be collected for analysis. Data from within-lesson self-assessment as well as end-of-unit criterion tests should be included. The same sort of analysis and the same standards of performance apply as those used in the earlier developmental testing cycle. Wherever fewer than 85 percent of the students meet the stated objectives, those portions of the systems that fall short may need some redesigning and revision. However, *all other possible sources of trouble should be examined before materials, methods, or media are changed,* because the system already has been thoroughly pre-tested during the development cycle. Special care should be taken to make sure that poor performance is not the result of installation problems, particularly systems management problems. If the system is being installed and tested at several different schools simultaneously, one of the potential sources of trouble could be some real differences in the various student populations. If the students are significantly different, the system will, of course, have to be revised to meet the local needs.

Neither the system designers nor the instructional staff should expect to implement a newly designed system without running into problems that will call for some revision. Local conditions will always affect the system to some degree. Therein lies the advantage of the systems concept—instructional systems signal the need for adaptation to local and changing needs. Feedback from continuous criterion performance testing provides a built-in mechanism for adaptive self-correction. *A training system is never a finished product; rather it is a continuing process for meeting the differing and changing needs of the individual student.*

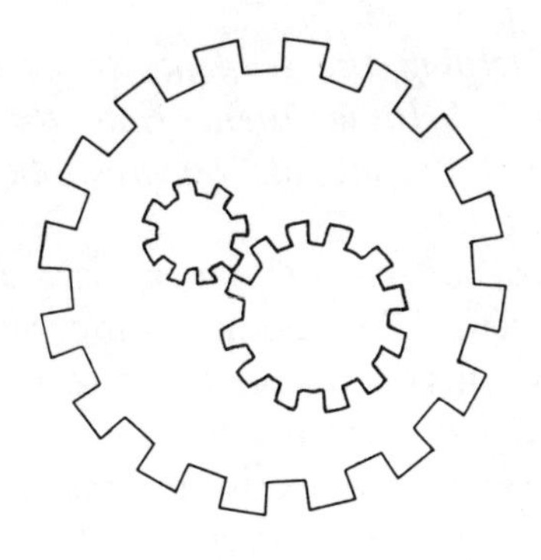

CHAPTER XV

ACCOUNTABILITY AND SYSTEM EVALUATION

Introduction

Providing the best possible occupational training program is only one of the major responsibilities of a vocational school. In addition, the school must provide a system for student career decision-making and job placement. Moreover, the school is accountable to the community for the quality and the utility of its product—the students it has trained for the job market. If vocational schools are to warrant continued (or even higher-level) financial support, they will have to demonstrate their effectiveness in terms of product success.

Guidance and Placement Services

Although the subject may seem to be beyond the scope of this discussion, effective guidance and placement services are so vital to the final success of an occupational education program that they should, in fact, be considered an integral part of the overall instructional system. The major reason for the inadequacies of many guidance and placement programs is that the schools, failing to recognize their importance, treat them as separate functions—apart from the educational process—and downgrade them accordingly. *Helping the students make the right occupational decisions and helping them find the right job after graduation are as much the schools' responsibility as providing the right training program.*

The overall goal of the guidance program is to prepare students in grades seven, eight, and nine for choosing a high school educational program. This occurs for the student during grade nine as a practical necessity. *Preparation for this choice should begin as early as possible,* since the decisions may be viewed as an important step leading to the achievement of later educational and occupational goals. An appropriate choice brings the student closer to the accomplishment of his goals with a minimum of penalty for error.

A "career decision" program should have carefully limited objectives. It should not attempt to satisfy all the needs of students normally identified as guidance functions. From a total domain of possible guidance functions and objectives, the program should deal only with those which specifically relate to educational and vocational decision-making. The activities are mainly organized to help the student acquire decision-making capabilities. *This decision-making process should be coordinated with other aspects of the existing guidance program*—incorporating the materials, projects, and programs already in use in the system.

The guidance program must be directly related to the overall curriculum design, and thus should draw heavily from the objectives and content of each training program. The career planning and decision-making process must be an integral part of the entire vocational development of the students, and not relegated to periodic conferences with a counselor. In fact, the program should be considered and developed as a prerequisite portion of the vocational training itself, with specific performance objectives and supporting individualized learning materials. Career decision-making is a continuing part of everyone's life-long occupational growth; thus, *the guidance program should serve as a model decision-making strategy that the student will find useful throughout his life.* (See Appendix M.)

To be effective, the guidance program must require each

student to be an active participant, not just the passive recipient of decisions handed down by a counselor. Making a decision implies action by the decision-maker. In fact, there is a series of actions and steps throughout the process. The individual making the decisions should engage directly and actively in all the steps needed to make choices and to adjust his decisions. The student has to be the principal agent in completing each step of the decision-making process. *The student, himself, must be given final responsibility for his decision.* He thus gets experience in using the basic tools of investigation and analysis and applying these to the decision-making process while developing his own career plan.

Decision-making is a continuous process which is influenced by all the changes which occur in the individual and in the environment of the individual. Vocational choice usually proceeds from fantasy through a mature decision-making process that requires reclassification and refinement of data at each step. The process of data collection, analysis, and summary can be accomplished at several levels, proceeding from the general to the specific. Therefore, the guidance program should call for *repetitive cycling of similar activities in the career decision-making process in grades seven, eight, and nine.* The steps through which the students proceed each year should remain the same, but the information and process should become increasingly more specific. Thus, each year the student learns to coordinate new facts with those he has already accumulated and to integrate them all in a coherent set of conclusions. Each year the students should repeat and review the following steps in the career decision-making program:

1. Setting goals.
2. Evaluating characteristics of self.
3. Identifying educational and vocational opportunities and requirements.
4. Selecting a course of action.
5. Assessing progress toward goals.
6. Adjusting decisions.

The main thrust of the guidance program must be to concentrate the decision-making activities on the students. However, few students would be able to arrive at sound career decisions without the help of their instructors and the special counseling staff. In these instances, the instructors and counselors should step in to help with the particular problem at hand, but only to help and not to make the decisions. The guidance program should be *a cooperative effort among the students, instructors, and the counseling staff.* The success of the program depends on such cooperation, *but the emphasis must always remain on student-initiated activity and decision-making.*

A school's responsibilities for its students does not end with their graduation. *The schools have the primary responsibility for placing each graduate in a job in the field for which he was trained.* In fact, the schools have that obligation to all students at every stage of their schooling; whether it is for part-time after-school jobs, for summer jobs, or even (or especially) to students who, for some reason, are leaving school for full-time employment before graduating. To serve this important role, the schools have to provide effective job placement services. *Every school, in effect, has to become a full-time employment agency.*

Although most schools recognize this obligation, there often is no organized effort to meet it. Studies show that much of the responsibility for the considerable number of graduates who do not find jobs in their field of training can be laid to the lack of adequate placement services within the schools. *The products of a training program, like all saleable products, require a specialized sales program.*

Regardless of the excellence of its vocational training, a school must have an effective placement program to insure that the graduates find work in the field for which they have been trained. To be effective, any such program must be an *organized effort of the entire instructional staff and the guidance staff, under the direction of at least one full-time placement coordinator.* Contact should be made with each potential graduate early in the year to assess his placement needs. A control chart should be maintained in the months prior to graduation to assure that special efforts will be made for those students yet to be placed. Contacts should also be made early in the year with potential employers to ascertain their future employment needs. Actually, such contacts should be only one part of a continuing program to improve the relationships between the school and potential employers in the area. *Only with a concerted and coordinated placement effort, can a school be sure that its students will find work in the field for which they were trained—the ultimate mark of success for a vocational training program.*

Product Follow-Up and Evaluation

Actually, the responsibility of a school does not end with the placement of its students, graduates or not, in training-related jobs. Although job placement is in itself an important measure of success, it is more important to determine the long-term career progression and life style of all former students. *The time has come to hold schools accountable for the success of their product.* For too long vocational education (and all education, for that matter) has been trying to justify itself by measuring the resources that go *into* the system rather than evaluating the product—the graduate. The majority of evaluation reports still concentrate on comparative studies of teacher salaries, equipment lists, and costs, student-staff ratios, physical plant costs, etc., paying *little or no attention to the qualitative success of the students after they have left the schools.*

In the very first paragraph of the introduction to this book, the systems process was described as involving *the setting of specific performance objectives and the subsequent rigorous measurement of the product against those performance objectives.* Almost the entire ensuing discussion of the instructional system development process and product has centered around those two essential elements. If, as described in Chapter I, vocational education is to become an instructional system capable of providing the social community with a predictable product, then *that product must be measured against the performance objectives set by the community.*

Therefore, a very important part of the job of the placement director and the guidance staff is to plan and carry out regular follow-up studies on all students (including actual dropouts) who have left school to enter the world of work. Such studies should chart the former students' careers after a lapse of one, three, five, and ten years. Appendix O includes some sample employer and student survey instruments that have been used for follow-up studies. The goal of the studies is to *provide the school with the evaluative feedback on product performance against which to judge past system performance and with which to plan for future system adaptation to the changing needs of the community and the students.* Any such study should be as much concerned with qualitative data as with purely quantitative findings. Job satisfaction and life style are at least as important as salary and job stability in determining the quality of the system's products. The introduction of this evaluative feedback into the system completes the regenerative cycle of the entire process. Follow-up evaluation of all students' performance in the world of work provides the only valid long-term assessment of system effectiveness. Like all productive systems, *an instructional system is judged finally by the quality of its product—by weighing output against input.*

APPENDIX A

SAMPLE JOB DESCRIPTIONS AND PERFORMANCE OBJECTIVES FOR GENERAL ELECTRONICS ASSEMBLER AND ELECTRONICS ASSEMBLY TESTER

GENERAL ELECTRONICS ASSEMBLER

A. *Defining the Population*

The majority of general electronics assemblers are employed by end-product and component manufacturing firms. Most assemblers are considered to be semi-skilled workers, although a significant number are placed on the skilled level. About 130 hours of training are required for workers employed in this area. This amounts to daily double periods (90 minutes) for one semester. Excluded from this definition are:

1. Wireworkers whose primary concern is making wire harnesses and wiring installations.
2. Production Repairmen whose primary mission is to locate and repair or replace defective components or wiring.
3. Developmental Assemblers whose primary mission is to assist higher-level personnel in developing new or prototype assemblies.

B. *Statement of Mission and General Objectives*

The primary missions of the general assembler are to:

1. Install wiring and components in electronic assemblies.
2. Perform minor testing of completed electronic assemblies.

C. *Duties*

The main steps in the occupation of the general assembler are to identify what components and wiring are needed to complete the assembly, install components and wiring and perform minor testing of completed assemblies.

D. *Contingencies and Contexts*

1. May have to substitute components with appropriate replacement parts.
2. May have to perform higher-level test procedures.
3. May remove or replace defective components.

In some situations the general assembler may work on only one type of assembly, while in smaller manufacturing establishments he may work on a variety of assemblies. In some situations, he will complete the assembly which was partially done by machines or other workers.

E. *Basic Tasks*

1. Identifies components.
2. Reads schematic diagrams.
3. Selects components.
4. Selects types of wires.
5. Solders wires and components.
6. Installs components (mechanical).
7. Makes wire harnesses (laces wires).
8. Selects hardware.
9. Uses hand tools.
10. Performs continuity checks.
11. Lays out assemblies.
12. Fabricates chassis.
13. Makes visual and mechanical inspections.
14. Replaces defective components and/or wiring.

Task List

1. Cut wire to prescribed lengths.
2. Demonstrate the various methods for stripping insulation.
3. Demonstrate the various methods for tinning wires.
4. Identify various electronic components and their respective symbols.
5. State the values of various resistors by using the RETMA Color Code.
6. Make a parts list from a given schematic.
7. State the circuit application of color-coded hook-up wire.
8. State the circuit application of color-coded transformer leads.

9. Lay out and prepare chassis for mounting components.
10. Mount components on chassis according to given specifications.
11. Terminate braid of shielded cable.
12. Make mechanical connections of wires to various terminal points.
13. Identify the various types of solder.
14. Identify the various types of soldering equipment.
15. Demonstrate proper use and care of soldering tools.
16. Solder wires to specified terminals.
17. Demonstrate proper unsoldering techniques.
18. Identify a printed circuit board.
19. Position and mount components on a printed circuit board in accordance with prescribed standards.
20. Solder components to printed circuit.
21. Demonstrate proper unsoldering techniques for printed circuit boards.
22. Demonstrate techniques for removal of components from printed circuit boards.
23. Connect component leads to specified terminals according to prescribed standards.
24. Solder component leads to terminal points.
25. Demonstrate the various techniques for lacing wires.
26. Route and solder a wire harness between terminal points according to prescribed specifications.
27. Make a continuity check of routed wires.
28. Identify conductors and insulators.
29. Relate the movement of electrons to current.
30. Define the following terms:
    current
    voltage
    resistance
    power
31. State the unit of measurement for:
    current
    voltage
    resistance
    power
32. Perform a continuity check.

Performance Objectives
General Electronics Assembler

1. Terminology
2. Manipulative Skills
3. Mechanical Measurement
4. Circuit Construction

1.0	*Terminology*	
1.1.1	Terminology	Component illustrations: names–matching
1.1.2	Terminology	Component illustrations–labeling
1.1.3	Terminology	Parts selection–given name and ratings
1.1.4	Terminology	Tubes–alphanumerical stocking
1.2.1	Terminology	Symbols: names of components–matching
1.2.2	Terminology	Schematic symbols–labeling
1.2.3	Terminology	Color Codes–reading
1.3.1	Terminology	Terms and definitions–matching
1.3.2	Terminology	Technical terms–sentence completion
1.3.3	Terminology	Component illustrations: characteristics or applications–matching
1.3.4	Terminology	Technical terms–writing definitions
1.4.1	Terminology	Pencil-and-paper exam–practice taking

1.1.1 Given illustrations of 30 widely used electronic components, and a randomized list of the names of the components, the student will be able to correctly match a minimum of 25 of the illustrations to the names of the components within ten minutes.

1.1.2 Given illustrations of 30 widely used electronic components, the student will be able to correctly label from recall and label a minimum of 25 components within ten minutes.

1.1.3 Given a description or designation of an electronic part, the student will be able to select the part from the stock of electronic parts. Electronic parts will include tubes, transistors, capacitors, switches, transformers, coils, fuses, wire, cables, and similar such parts. Examples of descriptions or designations are:

Tube — 6SN7
Resistor — 1.2 K-ohm, ½-watt, 5% tolerance, axial leads
Capacitor — .05mfd, tabular paper, 10% tolerance, 600-volt
Fuse — .25 amp, 250-volt, 1-inch
Math: Read decimals and units in various notation forms.

1.1.4 Given a group of 25 tubes, the student will be able to stock the tubes alphanumerically.
Math: Read and order by alphanumeric code.

1.2.1 Given drawings of 30 symbols widely used in electronic schematic diagrams, and a randomized list of the names of the components which they represent, the student will be able to correctly match a minimum of 25 of the symbols to the names of the components within ten minutes.
Math: Decimal, exponential, and prefix notation.

1.2.2 Given a list of 30 symbols widely used in electronic schematic diagrams, the student will be able to write from recall and label the names of the components the symbols represent.
NOTE: Assumes use of different symbols than for matching terminal objectives.

1.2.3 Given the following types of components or illustrations of the components, the student will be able to state the rating data indicated by the color code.

1. Radial lead resistor
2. Axial lead resistor
3. Molded choke coil
4. Molded tubular paper capacitor
5. Mylar/polyester film capacitor
6. Temperature compensating tubular ceramic capacitor
7. Standoff ceramic capacitor
8. Disc ceramic capacitor
9. Molded flat mica capacitor
10. Button silver mica capacitor

1.3.1 Given a list of 50 technical terms commonly used in electronics, and a randomized list of accurate definitions of the terms, the student will be able to correctly match a minimum of 45 technical terms to their definitions within 50 minutes.

Math: Notation (decimal, exponential prefix) and vocabulary (e.g., decade divider, pentode, triode, octal socket, etc.).

2.0 Manipulative Skills

2.1.1	Manipulative Skill	Soldering tip—installation
2.1.2	Manipulative Skill	Soldering tip—tinning
2.1.3	Manipulative Skill	Soldering tip—cleaning
2.1.4	Manipulative Skill	Insulation—stripping
2.1.5	Manipulative Skill	Solders spliced wires
2.1.6	Manipulative Skill	Solders conductors to terminals and connectors
2.1.7	Manipulative Skill	Solders components to terminals
2.2.1	Manipulative Skill	Heat sinks selection
2.2.2	Manipulative Skill	Heat sinks and soldering equipment—use to avoid heat damage
2.2.3	Manipulative Skill	Unsolders components from terminals
2.2.4	Manipulative Skill	Solders components to printed circuit board
2.2.5	Manipulative Skill	Unsolders components from printed circuit board
2.3.1	Manipulative Skill	Solders transistors to terminal strip
2.3.2	Manipulative Skill	Repairs break in wire coil
2.3.3	Manipulative Skill	Repairs break in printed circuit board conductor
2.3.4	Manipulative Skill	Soldering tools—selection for given task
2.4.1	Manipulative Skill	Safety procedures
2.5.1	Manipulative Skill	Constructs small electric motor

2.1.1 Given a soldering gun and a soldering tip, the student will be able to correctly install the tip in the gun assembly.

2.1.2 Given a soldering iron with a clean tip, the student will be able to correctly tin the soldering iron.

2.1.3 Given an oxidized and corroded soldering iron, the student will be able to correctly clean the iron in preparation for tinning.

2.1.4 Given a soldering task, wire stripping tool, and insulated conductors, the student will be able to correctly strip the insulation from the conductors as required for the particular task involved.

2.1.5 Given soldering tools, supplies, and wire samples, the student will be able to solder specific types of conductor splices.

2.1.6 Given a soldering iron, soldering gun, or soldering pencil, heat sinks, solder, and work pieces, the student will be able to correctly perform the following operation: Solder conductors to specified types of terminals and connectors.

2.1.7 Given a soldering iron, soldering gun, or soldering pencil, heat sinks, solder, and work pieces, the student will be able to correctly perform the following operation: Solder components to terminals.

2.2.1 Given a soldering task and an assortment of tools and heat sinks required for completing the task, the student will be able to select those heat sinks most appropriate for performing the task.

2.2.2 Given a soldering task, the student will be able to use those procedures and equipment which most effectively avoid damaging the insulation of the conductors due to excess heat.

2.2.3 Given a soldering iron, soldering gun, or soldering pencil, heat sinks, solder, and work pieces, the student will be able to correctly perform the following operation: Unsolder components from terminals.

2.2.4 Given a soldering iron, soldering gun, or soldering pencil, heat sinks, solder, and work pieces, the student will be able to correctly perform the following operation: Solder components to printed circuit board.

2.2.5 Given a soldering iron, soldering gun, or soldering pencil, heat sinks, solder, and work pieces, the student will be able to correctly perform the following operation: Unsolder components from printed circuit board.

2.3.1 Given a transistor and mounted terminal strip, the student will be able to solder the transistor to the terminal strip using a clamp-type heat sink.

2.3.2 Given an open coil of No. 6 enameled copper wire, the student will be able to repair the coil.

2.3.3 Given a printed circuit board in which a printed conductor has been cracked, the student will be able to repair the defect.

2.3.4 Given a soldering task and an assortment of tools required for performing the task, the student will be able to select those tools most appropriate for performing the task.

2.4.1 Given a list of tasks and a list of procedures to be followed in carrying out the tasks, the student will be able to correctly match the tasks with a procedure which accomplishes the task according to stated laboratory standards for safe practices.

2.5.1 Given a kit containing all the necessary parts for constructing a motor, the student will be able to correctly assemble the kit within one hour.

3.0 Mechanical Measurement

3.1.1 Measurement Ruler: English and metric
3.1.2 Measurement Drill gauge number and drill sizes
3.1.3 Measurement Thread gauge number and thread sizes
3.1.4 Measurement Wire gauge and wire diameters

3.1.5	Measurement	Micrometer: English and metric
3.1.6	Measurement	Vernier Calipers: English and metric vernier calipers
3.1.7	Measurement	Thermometers: Fahrenheit and Centigrade
3.1.8	Measurement	Weighing: English and metric scales

3.1.1 Given a ruler and work pieces, the student will be able to correctly use the ruler to make measurements of length in the English and metric systems.

3.1.2 Given a drill gauge and drills, the student will be able to measure the size of the drills.

3.1.3 Given a thread gauge and an assortment of machine bolts, the student will be able to use the thread gauge to measure the thread sizes of the machine bolts.

3.1.4 Given a wire gauge and an assortment of wire sizes, the student will be able to measure the diameter of the various wire samples.

3.1.5 Given a micrometer and an assortment of sheet metal samples, the student will be able to measure the thickness of the various sheet metal samples in the English and/or metric system.

3.1.6 Given a vernier caliper and an assortment of work pieces, the student will be able to measure the lengths of the work pieces in the English and/or metric system.

3.1.7 Given a thermometer and a source of unknown temperatures, the student will be able to measure the unknown temperatures in degrees Fahrenheit or centigrade.

3.1.8 Given a balance or scale and an assortment of unknown weights, the student will be able to measure the unknown weights in the English and metric systems.

4.0 *Circuit Construction*

4.1.1 Circuit Construction Terminates zip cord in plugs

4.1.2	Circuit Construction	Crimps terminals to wire
4.2.1	Circuit Construction	Tubes and transistors—selection from stock, given name
4.2.2	Circuit Construction	Tubes and transistor pins—location from schematic or manual
4.3.1	Circuit Construction	Kit of components and punched chassis—assembly
4.3.2	Circuit Construction	Chassis holes, given drilling tools
4.3.3	Circuit Construction	Drills and taps holes
4.3.4	Circuit Construction	Chassis mounting of components—drills or punches holes
4.4.1	Circuit Construction	Constructs a wire harness
4.4.2	Circuit Construction	Strings dial cords and springs
4.5.1	Circuit Construction	Simple radio kit—assembly

4.1.1 Given lengths of zip cord and two- and three-prong plugs, the student will be able to make the terminations using the Underwriter's approved knot.

4.1.2 Given an assortment of ten lugs and terminals, stranded and solid wire, a wire stripping tool and a crimping tool, the student will be able to correctly crimp the terminals to the wires within 15 minutes.

4.2.1 The student will be able to select a tube or transistor of a given type designation from a stock of 25 different tubes or transistors according to the name printed on the tube (or carton).

Math: Read alphanumeric codes.

4.2.2 The student will be able to locate the pin corresponding to a given tube or transistor element using a schematic diagram or tube/transistor manual.

Math: Geometry of reflection symmetry to determine clockwise or counter-clockwise counting for pin location.

4.3.1 Given a punched chassis, and a complete kit of electronic components, the student will be able to mount the parts and make all solder connections as required using a schematic diagram.

4.3.2 Given an electric drill, or drill press, work pieces, twist drill set, and other necessary tools (hammer, center punch, ream) the student will be able to select the correct twist drill and drill holes having the diameter required to provide a snug fit for a:

a. Potentiometer bushing
b. Rubber grommet
c. Plastic cord and strain relief

Math: Reading drill sizes and tables to select tool; measure diameters.

4.3.3 Given a set of six bolts of different size and thread, the student will determine correct tap size and thread for bolts, determine correct drill size for tap holes, and properly drill and tap a 1/8-inch aluminum plate for the bolts.

4.3.4 Given an electric drill, or drill press, work pieces, twist drill set, Greenlee punches and other necessary tools (hammer, center punch, ream) and hardware, the student will be able to mount the following items on a 1/16-inch thick aluminum or steel chassis:

a. Vacuum tube socket
b. Transistor socket
c. Variable capacitor
d. I.F. transformer
e. Power transformer

Math: Measurement and layout of measurements. Geometry of rectangular positioning of holes.

4.4.1 Given 10 conductors, cable ties, lacing cord, and a harness panel, the student will be able to construct a wire harness correctly dressed and laced as per a diagram within 15 minutes.

4.4.2 Given a dial kit, the student will correctly assemble dial, dial cords, and springs.

4.5.1 Given a simple radio kit (Greymark), the student will assemble the radio, following manual with a minimum of assistance.

ELECTRONICS ASSEMBLY TESTER

A. *Defining the Population*

The majority of electronics assembly testers are employed by end-product and component manufacturing firms. Most assembly testers are considered to be skilled workers, although some may be classified as only semi-skilled. About 130 hours of training (in addition to General Assembly training) are needed for workers employed in this area. Excluded from this definition are:

1. Systems Testers who test complete electronic systems such as radio or television transmitters and computer memory units.
2. Testing and Regulating Technician who tests and regulates installed telephone and telegraph equipment and terminal to maintain continuity of service.

B. *Statement of Mission and General Objectives*

The primary missions of the assembly tester are to:

1. Adjust tuned circuits in radio and television receivers using standard and special test equipment.
2. Test assembled electronic devices for compliance with specifications using standard and special test equipment.
3. Test assembled electronic components with oscilloscope and multimeter to detect missing parts, loose wires and defective solder joints.
4. Disassemble electronic devices and components and repair or replace defective parts.

C. *Duties*

The main steps in the occupation of assembly tester are to

determine if the electronic device or component performs according to specifications and, if not, to locate and eliminate the source of failure by re-aligning, repairing or replacing defective parts.

D. *Contingencies and Contexts*

1. May have to substitute components with alternative replacements.
2. May have to perform some fairly complex trouble shooting tasks.
3. May have to instruct assemblers in proper techniques to eliminate recurring malfunctions in assembled components.
4. May have to develop new assembly techniques to eliminate production problems.

E. *Basic Tasks*

1. Identify function of test equipment including multimeters, vacuum tube voltmeters, audio generators, signal generators, power supplies, oscilloscopes, etc.
2. Set up and adjust all the above test equipment.
3. Read meter scales.
4. Read trace patterns on oscilloscope.
5. Determine test points and operating values from wiring diagrams and schematics.
6. Perform test procedures.
7. Apply basic laws of a.c.-d.c. electricity to the detection of opens and shorts.
8. Apply knowledge of construction and function of basic electronic components to the detection of faulty components.

Task List

1. Identify oscilloscope.
2. State use of oscilloscope.
3. Identify multimeter(s).
4. State the use of the multimeter.
5. Name the controls and specify their functions.

6. Make proper preliminary adjustments.
7. State safety precautions.
8. Identify meter dial (scales) information.
9. Read and interpolate a meter scale.
10. Demonstrate use of multiplier.
11. Demonstrate use of meter function switch.
12. Demonstrate the use of the range selector.
13. Identify gradicule scale.
14. Read and interpolate gradicule scale.
15. Demonstrate use of multiplier.
16. Demonstrate use of sync control.
17. Calibrate, using the gradicule scale.
18. Identify test point on a wiring diagram.
19. Identify same test point on equipment.
20. Identify test point on a schematic.
21. Identify same test point on equipment.
22. Identify electrical abbreviations used on wiring diagrams and schematics.
23. Ready the measuring equipment.
24. Make connections for performing tests.
25. Perform tests.
26. Record results.
27. Tag tested equipment.
28. State the methods of producing electricity.
29. State the relationships among current, voltage, power, and resistance in a series circuit.
30. State the relationships among current, voltage, power, and resistance in a parallel circuit.
31. State the relationships among current, voltage, power, and resistance in a series parallel circuit.
32. Measure the voltage, current, and resistance in the series circuit.
33. Measure the voltage, current, and resistance in the parallel circuit.
34. Measure the voltage, current, and resistance in the series parallel circuit.
35. Define an open circuit and a short circuit.
36. Define the terms: potential, ground, and reference point.
37. Locate an open in a simple 4-component circuit.
38. Locate a short in a simple 4-component circuit.

39. State the differences between a.c. and d.c.
40. Identify an a.c. and a d.c. voltage with a test instrument.
41. State two sources of a.c.
42. Define frequency and magnitude.
43. Measure and convert a.c. voltages.
44. Draw waveforms.
45. Describe construction of an inductor.
46. State function of an inductor.
47. State unit of measure of inductance.
48. Describe construction of a capacitor.
49. State function of a capacitor.
50. State unit of measure of a capacitor.

Performance Objectives
Electronics Assembly Tester

5. Electrical Measurement
6. Circuit Design
7. Circuit Analysis
8. Trouble Shooting

5.0	*Electrical Measurement*	
5.1.1	Measurement	Reads printed illustrations of meter scales
5.1.2	Measurement	Reads scales on various instruments
5.2.1	Measurement	Battery terminals: identifies positive and negative
5.2.2	Measurement	Battery voltages, terminal-load vs. no-load
5.2.3	Measurement	Battery, specific gravity reading
5.2.4	Measurement	Battery condition from performance measurements
5.3.1	Measurement	Multimeter—selects scale and range for practical task
5.3.2	Measurement	Multimeter—applies meter to circuit
5.3.3	Measurement	Multimeter, current, voltage, and resistance readings
5.3.4	Measurement	Multimeter readings—records data
5.4.1	Measurement	Waveform on oscilloscope—displaying
5.4.2	Measurement	Calibrates scope with internal signal
5.4.3	Measurement	Voltage amplitude of unknown signal with

		scope—measurement
5.4.4	Measurement	Sine wave values—conversion between max., effective peak to peak
5.5.1	Measurement	Construct simple volt-ohmeter (Greymark kit)

5.1.1 Given illustrations of meter faces, the student will be able to read pointer indications on the following types of scales to a specified accuracy.*

a. Linear
b. Logarithmic
c. Exponential
d. Scales a, b, or c, with both left- and right-hand zero
e. Scales with center zero
f. Decibel scales

*Scales should be read to the nearest whole, half, or tenth of a division.

5.1.2 The student will be able to read the scales on the following types of instruments:

a. Micrometer (English and metric)
b. Vernier dial
c. Watt hour meter

5.1.3 The student will be able to test a given tube or transistor in a commercial laboratory-type checker for all parameters which the instrument is capable of checking. The student will check 20 out of 25 tubes and transistors in 20 minutes.

5.2.1 Given an assortment of batteries, the student will be able to identify the positive and negative terminals.

5.2.2 Given the following batteries, a voltmeter and other necessary components, the student will be able to measure the terminal voltage of the batteries under no-load and rated full-load conditions.

A. 1.5v (size D) dry cell
B. 1.5v (size C) dry cell

C. 5v dry cell
D. 9v dry cell
E. 22v dry cell
F. 30v dry cell
G. 45v dry cell
H. 90v dry cell

5.2.3 Given a storage battery and a hydrometer, the student will be able to measure the specific gravity of the battery electrolyte.

5.2.4 Given a set of batteries that vary in condition, the student will judge condition of each battery based on his performance measurements.

5.3.1 Given a Volt-ohm-ammeter and a source of unknown a.c. and d.c. voltages, currents, and resistances, the student will be able to select the correct scale and range required to obtain the reading.
Math: Scale readings (selection of correct number of significant figures, rounding off numbers.)

5.3.2 Given a Volt-ohm-ammeter and a source of unknown a.c. and d.c. voltages, currents, and resistances, the student will be able to correctly apply the instrument to a circuit in order to read a voltage, current, or resistance.

5.3.3 Given a Volt-ohm-ammeter and a source of unknown a.c. and d.c. voltages, currents, and resistances, the student will be able to use reading procedures (avoid parallax, interpolate values, etc.) which provide accurate and consistent readings.

5.3.4 Given a Volt-ohm-ammeter and a source of unknown a.c. and d.c. voltages, currents, and resistances, the student will be able to use data recording procedures (use significant figure and rounding off procedures, etc.) which provide clear and unambiguous statements of the magnitude and uncertainty in the readings.

5.4.1 Given an oscilloscope and an a.c. signal of unknown frequency and amplitude, the student will be able to make the necessary adjustments to display two complete cycles of the waveform on the screen.
Math: Grid counting, ratio of units, division (proper choice of reference voltage). Convert r.m.s. voltage value to peak voltage.

5.4.2 Given an oscilloscope and an a.c. signal of unknown frequency and amplitude, the student will be able to make the necessary adjustments to calibrate the scope using the internal calibration signal.

5.4.3 Given an oscilloscope and an a.c. signal of unknown frequency and amplitude, the student will be able to make the necessary adjustments to measure the amplitude of the unknown signal in volts with a maximum ± 5 percent error.
The student will be able to make 9 correct readings out of 10 determinations of each quantity in 10 minutes.

5.4.4 Given the value of any one of the sine wave parameters, average, r.m.s., peak, or peak-to-peak, and appropriate conversion formulas, the student will be able to convert the given value to the value of other parameters.

5.5.1 Constructs a simple volt-ohmeter from kit provided (Greymark), using manual of instruction, with little or no help from others.

6.0 Circuit Design

6.1.1 Circuit Design Component to meet given specifications—selecting
6.2.1 Circuit Design Determines rated maximum current capacity of given wire size and vice versa
6.2.2 Circuit Design Wire resistance—computing
6.3.1 Circuit Design Connects batteries to obtain required voltage and current
6.3.2 Circuit Design Determines internal resistance of battery
6.3.3 Circuit Design Power equation, one unknown—solving

		for unknown
6.3.4	Circuit Design	Equivalent resistance—computing
6.3.5	Circuit Design	Resistances for voltage divider—specification
6.4.1	Circuit Design	Determines and specifies load resistance for maximum power into load
6.4.2	Circuit Design	Calculates fuse needed for given circuit
6.4.3	Circuit Design	Determines ballast resistor value
6.5.1	Circuit Design	Connects components for given switch positions
6.5.2	Circuit Design	Draws wiring and schematic diagram for given specifications

6.1.1 Given the necessary specifications—i.e., voltage, current, wattage, etc.—the student will be able to select a component meeting these specifications from a commercial catalog of electronic parts.
Math: Reading decimals and units.

6.2.1 Given a list of wire sizes and a table of conductor current testing data, the student will be able to determine the current carrying capacity of the wire; or conversely, given the required current carrying capacity of a conductor, the student will be able to determine the necessary wire size.
Math: Reading decimals, reading tables.

6.2.2 Given a specified wire size and a table of wire resistances, the student will be able to compute the resistance of a specified length of the conductor.

6.3.1 Given a number of cells of various voltage and current ratings, the student will be able to connect these to obtain a specified voltage or current capacity.
Math: Logic and adding and subtracting decimal values of current and voltages.

6.3.2 Given a battery and VOM, the student will be able to determine and state the internal resistance of the battery.
Math: Reading decimal values. *Ohm's Law.*

6.3.3 Given any two of the factors in the power formulas:

$$P = I^2R, \frac{E^2}{R}, \text{ or } EI,$$

the student will be able to solve for the unknown factor.

6.3.4 Given any number of specified resistances in parallel or in series, the student will be able to solve for the value of equivalent total resistance.
Note: Verify by lab measurement.

6.3.5 Given the voltage specifications for an unloaded voltage divider, consisting of approximately three resistors and a battery in series, the student will be able to determine the value of resistances required.

6.4.1 Given a signal or power source, the student will be able to determine and specify the value of load resistance required for maximum power transfer to the load.
Math: Use of Ohm's Law and Power Law (P = EI).

6.4.2 Given a complete circuit containing up to six bulbs and/or other devices, the student will be able to calculate the size of fuse required to protect the circuit from overload.
Math: Use of Power Law (P = EI), and adding decimal values of currents.

6.4.3 Given a lamp or vacuum tube type and catalog, the student will be able to calculate the value of ballast resistor required to operate the lamp or tube filament from a voltage other than the rated voltage.
Math: Use of Ohm's Law. Addition and subtraction of reciprocal values of resistance:

$$\frac{1}{R_T} = \frac{1}{R_1} + \frac{1}{R_2} + \frac{1}{R_3} \cdots$$

6.5.1 Given a battery, switches, jumpers, and lamps, the student will be able to connect these components so that a specific lamp(s) will light when using designated switch positions.
Math: Logic (If . . . , then . . .)

6.5.2 Given a circuit performance specification (e.g., student is told that certain bulbs will be operated under specified voltage, current conditions, and specified switching conditions, etc.), the student will be able to draw a wiring diagram and schematic diagram for a circuit which properly connects one or more lamps in series, parallel or series parallel arrangements.
Math: Logic (Ohm's Law).

7.0 Circuit Analysis

7.1.1	Circuit Analysis	Schematic of pictured circuits—drawing
7.2.1	Circuit Analysis	Unknown voltage in a loop—solution
7.2.2	Circuit Analysis	Unknown current—specifies current and direction
7.3.1	Circuit Analysis	I,E,R—specifies effect of change in one
7.3.2	Circuit Analysis	I,E, or R (Ohm's Law)—calculation
7.3.3	Circuit Analysis	P,E,I,R (Power Law)—calculation
7.3.4	Circuit Analysis	Parts list from schematic diagram—writing
7.3.5	Circuit Analysis	Bulb problem: which will be brightest
7.4.1	Circuit Analysis	Circuit behavior—selects statements which will describe

7.1.1 Given a simple working electronic circuit or photograph, the student will be able to draw a schematic diagram of the model.
Math: Decimals needed in labeling.

7.2.1 Given all of the voltages and voltage drops in a circuit loop, except one, the student will be able to specify the value of the unknown voltage.
Math: Addition and subtraction of decimal voltage values.

7.2.2 Given the value of all of the currents into and out of a circuit junction, except one, the student will be able to specify the value and direction of the unknown current.
Math: Addition and subtraction of decimal current values; conversion between verbal notation and numerical notation, e.g., add 20 m.a. to .05 amps.

20 m.a. = .02 amps

$$\text{so}\quad \begin{array}{r} .05 \\ \underline{.02} \\ .07 \end{array} \text{ amps or 70 m.a.}$$

7.3.1 Given the Ohm's Law formula:

$$I = \frac{E}{R},$$

and a list of statements describing the relationship between each of the factors I, E, and R, the student will be able to select statements which correctly describe how any one factor will respond to changes in any other factor, the remaining factor being held constant.
Math: Multiplication and division of decimals. Units kept in order.

7.3.2 Given any two of the factors I, E, or R, the student will be able to solve for the remaining unknown factor in the Ohm's Law formula:

$$I = \frac{E}{R}.$$

A. Given E and I, find R.
B. Given R and I, find E.
C. Given E and R, find I.

7.3.3 Given any two of the factors E,I, and R, and given the set of formulas, the student will be able to solve for the power P in the following formulas:
(a) $P = E^2/R$
(b) $P = I^2R$
(c) $P = EI$
Math: Squaring, taking square roots (from table), multiplying and dividing decimals. Keeping units in order.

7.3.4 Given a schematic diagram, the student will be able to write a complete parts list for the device represented by the diagram.

7.3.5 Given a circuit containing up to six bulbs, the student will be able to state which one of the bulbs will light the brightest.

7.4.1 Given 8 schematic diagrams of circuits employing a battery, 2 single-pole, single-throw switches, a maximum of four lamps, and a list of twelve statements which describe the operation of the circuits under various switching conditions, select the four statements required to completely describe the behavior of each circuit under all switching conditions. The student should be able to obtain at least 25 out of the 32 possible answers correctly within 30 minutes.

8.0 Trouble Shooting

8.1.1	Trouble Shooting	Oscilloscope waveform drawings–matching
8.2.1	Trouble Shooting	Components: possible defects–matching
8.2.2	Trouble Shooting	Defective components: detection tests–matching
8.3.1	Trouble Shooting	Defective lamp–location
8.3.2	Trouble Shooting	Defective component: anomalous test voltage–locate component
8.3.3	Trouble Shooting	Malfunction–diagnosis and checking, tags solution
8.3.4	Trouble Shooting	Malfunction–diagnosis and specification of cause and site
8.3.5	Trouble Shooting	Malfunction–selection and use of repair tools, repair
8.4.1	Trouble Shooting	Circuits and diagrams–correction
8.4.2	Trouble Shooting	Sounds: sources–matching
8.4.3	Trouble Shooting	Abnormal operation–diagnosis by sounds
8.5.1	Trouble Shooting	Split-half search technique

8.1.1 When shown drawings of pairs of oscilloscope waveforms, the student will be able to identify pairs which are the same and pairs which are different.

8.2.1 Given a list of components, and a randomized list of defects, the student will be able to correctly match the ways in which the component may become defective to the names of the components.

8.2.2 Given a list of components and their defects, and a list of test procedures, the student will be able to correctly match the defective component to a test procedure appropriate for detecting the nature of the defect.
Need T.O. for sequence?—i.e., student trouble shoots in proper order.

8.3.1 Given the circuit shown below in which a string of 12 lamps are connected in series and in which any one of the lamps may be defective, the student will be able to locate the defective lamp by using an ohmmeter and removing no more than 3 lamps in order to make continuity tests.
Math: Logic, reading.

8.3.2 Given a schematic diagram with normal voltages appearing at key test points, the student will be able to specify which defective component(s) could cause a higher or lower than normal voltage at a designated point.
Math: Read and compare decimal voltages.
Math logic (Ohm's law).

8.3.3 Given electronic circuits of the type to be illustrated and schematic diagrams defining test points and test point values and tolerances, the student will be able to:

Record the initial state of a model as a normal or malfunctioning model; designate the malfunctioning stage, the specific cause of the malfunction, and the corrective action taken, by checking the corresponding boxes on a tag attached to the model.

8.3.4 Given simple electronic circuits and schematic diagrams defining test points and test point values and tolerances, the student will be able to:

Select and correctly use the appropriate test

equipment (1) to determine whether or not a model is functioning within the "normal" ranges specified for its best point values and, (2) if the model is not functioning within the "normal" ranges, determine the malfunctioning stage and the specific cause of the malfunction.

Math: Reading linear, log, and exponential decibel scales to intermarker values and writing these readings. Interpreting readings in relation to given tolerances; e.g., 85 is satisfactory if rating is 100 ± 20%.

8.3.5 Given simple electronic circuits and schematic diagrams defining test points and test point values and tolerances, the student will be able to select and correctly use the appropriate tools and materials required to repair the malfunction.

8.4.1 Given an incorrectly wired circuit board and a wiring diagram of the circuit, the student will be able to correctly rewire the circuit.

8.4.2 When presented recordings of various sounds useful in the detection of the normal or abnormal operations of electronic equipment, and a list of terms (such as hum, buzz, and boom), the student will be able to match the terms to the names of the equipment which usually generates the sound.

8.4.3 Given a list of components or devices and a randomized list of sounds indicating normal or abnormal operation, the student will be able to correctly match the component to the sound and specify whether normal or abnormal operation is indicated.

8.5.1 Given 3 different circuit boards, each combining series, parallel, and series parallel circuits and power supply, the student will locate the short or open in each; the student will demonstrate the "split-half" trouble shooting technique by using the minimum number of steps to locate each malfunction.

APPENDIX B

CRITERION TEST CONSTRUCTION

Knowledge Testing

There are basically three types of criterion test items that can be used to test knowledge.

1. A test item may be directive or imperative: *Find the value of R in an electrical circuit if I = 30 amperes and E = 110 volts.*
2. An item may be a completion type, requiring the student to select from several possible choices the one that he thinks will correctly complete the stem of the item: *Excessive backlash in the differential assembly of an automobile would most likely be caused by . . .* (followed by a blank space or four alternative responses).
3. An item may ask a direct question: *What three types of meter functions are combined in a multimeter?*

The three basic types of criterion test items can be adapted to measure all the different kinds of acquired knowledge. Remember, instructional objectives state the *training goals in performance terms*—what the student is to do to demonstrate that he has learned. Therefore, it is important to measure the students' knowledge by testing their ability to apply that knowledge to the problems they will encounter on the job, rather than by the mere recall of isolated facts.

One strongly recommended method is to provide hypothetical situations and then to ask practical, objective questions about courses of action that should be taken in the situations. The following sample test item is an example of a realistic applied mathematics problem.

An opening 6 yards long and 3 feet wide is to be covered with sheathing. Enough lumber is available to cover two-thirds of the area of the opening. How many more square feet will he have to get to finish the job? (a) 3; (b) 6; (c) 12; (d) 18.

Most well stated training objectives are easily translated to criterion test items, although considerable ingenuity may be needed in some instances. One goal is to develop test items that can be scored objectively—without scorer bias. The following paragraphs offer suggestions on how to construct several types of objective test items and also recommendations on when to use each.

Multiple-Choice Items

The multiple-choice type of question is preferred for most simple knowledge tests. It can also be used to test certain skills. Multiple-choice questions are versatile and can be used to measure knowledge of facts, terminology, concepts, principles, and applications. This type of question not only tests the ability to recognize but also the ability to discriminate among several alternatives. It is especially effective as a means of requiring the student to select what is relevant from an array of data.

In a multiple-choice question, the stem is a question or an incomplete statement. The stem is followed by four alternatives—possible answers or completions of the stem. The student has to select the best of these—the correct, most nearly correct, most comprehensive, etc., alternative of the four presented.

Here are several examples:

A waffle iron with a resistance of 20 ohms is connected to a 110-volt circuit. The current flowing in the circuit is: (a) .5 amperes; (b) 5.5 amperes; (c) 22 amperes; (d) 220 amperes.

The best tool for shaping a curved surface is the: (a) plane; (b) chisel; (c) spokeshave; (d) knife.

Radiating ribs are placed on automotive brake drums to: (a) balance the drums; (b) strengthen the drums; (c) cool the drums; (c) decrease drum wear.

Matching Items

Matching items are particularly appropriate to testing knowledge of materials that are factual and homogenous in nature. This form of presenting items uses two columns of related words, phrases, symbols or illustrations. The student then has to match each element in one list with the element in the other list to which it is most closely related. The number of alternatives in one list should always exceed the number of elements in the other by two or three to insure that there are always alternatives left for the last element.

Here is an example of a matching criterion test item:

Match each term listed below with the statement that best describes it.

1.	*The unit of electrical resistance.*	*a.*	*ampere*
		b.	*conductor*
2.	*A device used to open or close a circuit.*	*c.*	*electron*
3.	*The unit of electromotive force.*	*d.*	*insulator*
		e.	*joule*
4.	*The rate of flow of electrons in a circuit.*	*f.*	*ohm*
		g.	*switch*
5.	*The basic unit of charge.*	*h.*	*volt*

True-False Items

While true-false test items find their way into many tests, they have little or no place in realistic criterion tests. Supposedly, these items are well adapted to testing recognition of technical information and principles. However, present-day thinking is that the student may accumulate misinformation if he is deliberately presented with false information. Another drawback is that the test writer more often than not rewrites verbatim statements from the training materials. Sometimes negative terms are included to make some of the items false. The result is to encourage rote memorization.

Another reason for not using true-false items is that it is difficult to include in the item itself any standard by which the student can judge whether the item is true or false. To increase the validity of this type of item, it is often necessary to resort to statements that are unquestionably true or false. This again leads to the criticism that true-false items place too much emphasis upon memorizing facts. Moreover, the student always has a 50/50 chance of getting the right answer, fairly good odds if he is just guessing. It is recommended that true-false items not be used in criterion tests.

Completion Items

Completion items can also be used in a criterion test if they require very short answers that are limited to only a few words which can be scored objectively. Directions for such items should clearly indicate that the answer is to be limited to one, two, three, or four words.

Completion items usually take one of two forms. They can be questions requiring a single word or phrase as the answer, or they can be sentences with one or more blanks to be filled in with a word or short phrase. Some test writers have carried this to an extreme and have produced items that look like Swiss Cheese.

The _______ of electrons in a _______ can be stated in terms of _______.

_______ is the rate of _______ per unit of _______.

Completion items should test recall, but the student would have had to memorize the lesson material to call up the correct answers with the few clues presented in the two items above. Well written completion items, however, do have a place in criterion testing programs.

Essay Items

An essay item usually calls for extensive discourse by the student. The major difference between essay items and short-answer items is in the length of the response rather than in the wording of the items. The essay item is not appropriate for measuring simple recall. This type of question should be used only when the student is expected to do reflective or creative thinking, organize knowledge in the solution of a problem, and express his solution in writing.

The major disadvantage of essay items is that they cannot be scored objectively. Instructors will not agree with each other as to the completeness or accuracy of the answers. Essay items appear easy to write but they are not. If you want to test more than memory of facts, essay items are among the most difficult items to write. Poor essay items—such as, "Write all you know about the automobile engine"—are easy to write, but are obviously going to be difficult to score because of lack of explicitness. Thus, the essay item often turns out to be less satisfactory than a series of multiple-choice items to test the same objective.

Because the goal of criterion testing is to determine whether the student can meet the requirements of the training objective and not whether he can write extensively and well, essay items have little or no place in a criterion test. This fact, coupled with need to score the tests objectively, should lead you to reject essay items in most cases.

Short-Answer Items

The term "short-answer" means different things to different people. As used here, a short-answer item is a "supply-the-answer" kind of item that stands somewhere between the completion and essay question. It can be either a question or an imperative statement. The answer the student has to furnish is limited to a brief list, a phrase, or a short statement. Short-answer items can be used whenever the student is to recall facts, basic ideas or principles, or to make simple applications of what he has learned.

There is one prime rule to follow when writing this type of item. Be sure to limit the length of the possible answer. Make sure the student will know precisely what is wanted; otherwise, the answer will end up being a hard-to-grade essay. The single point being tested must be strongly emphasized and delimited.

A question like this one should not be asked:

> *Compare the methods of drying manufactured machine parts.*

Instead, ask him to:

Give two advantages of drying manufactured machine parts with compressed air.

The first item is obviously too broad. A student could write a long essay as his answer. On the other hand, the second item sets some limits and so the student knows precisely what is required. Here is another example of a poor short-answer item:

How do you cope with moisture in a refrigeration system?

Now look at a better version of the same question.

Why is silica gel packed with a new refrigeration system?

A student could go on at length in answering the first question while the second one calls for a limited and short answer:

To absorb moisture.

Graphic Items

Graphic items have an important place in criterion testing. This kind of item can present the "given" element pictorially and require either a graphic or a verbal response, or a combination of the two. Often the graphic item will come closer to testing the behavior defined in the objective than can a completely verbal item.

Here are two examples of graphic test items:

(1) The measurement indicated by the arrow is: (a) 11/16; (b) 9/16; (c) 5/8; (d) 3/4.

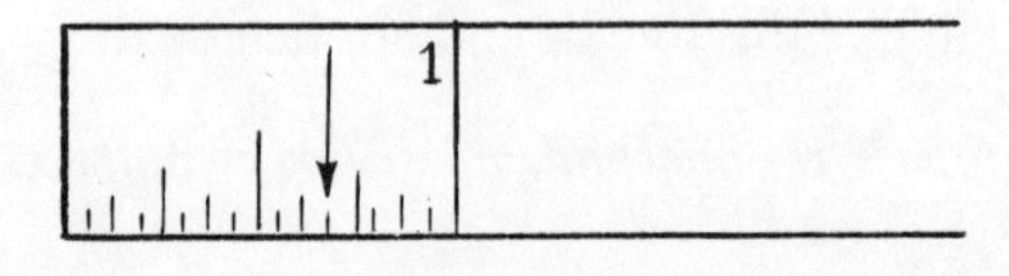

(2) *Each drafting term below describes one of the sketches. Match each term to the appropriate sketch.*

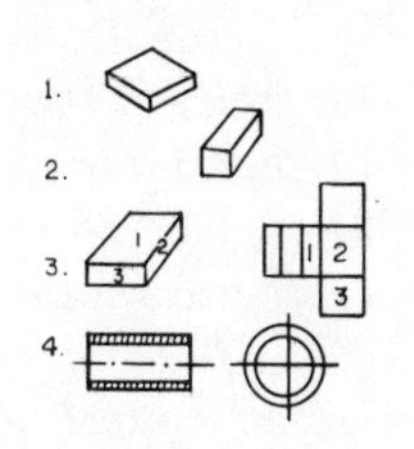

a. Full section.
b. Isometric projection
c. Oblique projection
d. Orthographic projection
e. Parallel-line development
f. Perspective
g. Revolved section

Problem-Solving Items

With problem items, the knowledge test is approaching a skills test. Criterion test items of the problem variety can be used for both knowledge and skills tests. This is highly desirable, since problem items are truly functional—they test the ability to apply what has been learned to new situations. Problem items require the student to apply rules to arrive at solutions, both mathematical and non-mathematical. To solve them, the student must understand concepts and principles and, more importantly, must apply them. The mathematical problem forces the student to perform operations accurately and in a definite sequence.

There are several categories of mathematical problems ranging from simple arithmetic to formula and equation problems. Here are a few examples:

Divide: 40 into 1,208

Answers: *a. 3*
b. 30
c. 33
d. None of the above

Two parallel pipes open at one end. They are to be joined by connecting an elbow to each and connecting the elbows with a new length of pipe. The center-to-center distance between the parallel pipes is 11 feet 3 inches. If the centerline of one of the parallel pipes is extended into the elbow, the distance from the extend-

ed centerline to the outer rim of the elbow where the new pipe will be joined is 1 foot 5/16 inches. The length of the effective thread of the pipe where it enters the elbow is 1 inch. How long should the pipe be cut?

a. 11 feet 5/8 inch
b. 11 feet 1 1/8 inches
c. 11 feet 1 7/8 inches
d. 11 feet 3 inches

A flashlight lamp is connected to a 1 1/2-volt dry cell and uses .1 ampere of current. What is the resistance of the lamp?

a. .015 ohms
b. .15 ohms
c. 1.5 ohms
d. 15 ohms

The following examples will assist in preparing criterion test items for several types of specific behavior. The list is by no means exhaustive, but it should help to construct test items that test specific kinds of learned performance. Many of the examples could be constructed as more than one type of test item.

1. Computes
   What is $n^3 + 4$ *equal to if* $n = 4$*?*
   How many tens are there in 378?

2. Calculates
   Divide: 27 into 4379.4

A circular saw cuts 8 boards per minute. If there are 1,440 boards to be cut, the number of hours required to cut the boards is:

a. 2 1/3
b. 2 2/3
c. 3
d. 4

3. Analyzes
   Refer to the schematic diagram of the washing machine and tell what would happen if fuse F-4 opened.

   If your contact print is too dark after 1 1/2 minutes of developing time, what should you do to the next print from the same negative?

4. Recalls
   List the types of flooring used in commercial buildings.

   What are the cores of electromagnets usually made from?

5. Classifies
   On which machine would each of the following shop functions be performed?

   - *a. Saw board to length.*
   - *b. Turn stock to 3-inch diameter.*
   - *c. Reduce thickness of rough boards.*
   - *d. Cut 4 x 8-foot plywood in half.*

   Classify the following items into two categories: raw materials and tools of the trade.

6. Itemizes
   Itemize the charges for long distance calls on the attached bill.

   List all the items you will need to take with you when you serve on the fire line.

7. Defines
   Give the meaning of the term "thermite welding" as used in automobile body shops.

Define "load" as used in electrical circuits.

8. Describes
What is backlash in a differential assembly?

Describe the correct way to connect an ammeter into an electrical circuit.

9. Identifies
List the common elements in trouble shooting a defective washing machine and a defective electric frying pan.

Which of the following addressed mail should have the same first digit in the zip code?

10. Compares
Rank the various types of gasoline engines according to their efficiency.

Is there any difference in the installation of sub-flooring and finish flooring?

11. Discriminates
Match the terms in the right-hand column with the statements on the left.

1.	*Machinist hammer*	*a.*	*Ball peen*
		b.	*Dividers*
2.	*A process which results in a finished surface*	*c.*	*Double cut*
		d.	*Draw filing*
		e.	*Sal ammoniac*
3.	*Used in layout of circles*	*f.*	*Single cut*
		g.	*Single hem*
4.	*Used to identify tool steel*	*h.*	*Spark test*
5.	*Quick-cutting file*		

In pipe fitting, a die is used for:

a. *Threading.*
b. *Planishing.*
c. *Brazing.*
d. *Cleaning.*

12. Explains
In 50 words or less explain:

How water can be removed from an air-brake system in a truck.

The correct procedure for dimensioning a drawing of a house foundation.

APPENDIX C

CONSTRUCTING MULTIPLE-CHOICE AND COMPLETION TEST ITEMS

Because most training courses will rely considerably on objective tests for testing knowledge, this section contains several suggestions for improving the quality of objective test items, especially multiple-choice items. For the most part, these suggestions also apply to other types of test items.

Writing effective test items is one of the most difficult tasks in developing a training system. Besides requiring considerable time and effort, item writing demands mastery of the subject, an ability to write clearly and concisely, and an ability to visualize job-like situations to include in the various problems.

A multiple-choice item presents a problem either as an incomplete sentence or as a question. The student is to choose the best answer to the question or the alternative that best completes the statement.

For clarity and simplicity, the direct question should be used most often. However, the incomplete sentence is sometimes more economical in the use of words. The goal should be to develop items in whatever form results in the least verbal difficulty for students. Do not add to their reading problem; if anything, minimize it.

The test should directly reflect the training objectives. The student should be required to perform the task exactly as required by the objectives. The test should *measure the ability of the student to perform the task,* not just his knowledge about the task.

Here are some suggestions for writing test items and some examples of poor and good items.

Dangling Constructions

Every test item should be expressed in clear, concise language. Awkward word arrangement and unnecessarily complex statements should be avoided. One common error is the use of dangling participles or gerunds, leaving the subject implied rather than stated. For example: *When an automobile is equipped with a ball-joint suspension system, the main advantage is . . .* This item can be improved by saying: *The main advantage of a ball-joint suspension system is . . .*

Another source of weak sentence structure is the use of "it" to start the item. For example: *It is the main advantage of an alternator system that it . . .* Compare the weak structure with the revised version: *The main advantage of an alternator system is . . .*

Negative Form

Items should be written in a positive form rather than negative. Unless there is very good reason to include items that ask the student to determine what is the *least* desirable, what is *never* a factor, or what is *not* a characteristic, most students prefer to indicate the *best* answer rather than the unacceptable one. Also, negative items invite trouble in the form of double negatives. Sometimes, however, the negative forms have to be used because more plausible wrong alternatives can be found. Poor example: *Which of the following is not an advantage of a transistorized ignition system?*

Alternative Answers

All the alternatives to the question or statement should be plausible and grammatically correct. If the alternatives do not seem to be reasonable answers to the item, they are not serving their intended purpose. All alternatives should be grammatically parallel. The best check is to reread the complete item with each alternative in turn to be sure each one follows the lead grammatically. Unless this suggestion is followed, many students will accept or reject an alternative because it is dissimilar from the other alternatives.

Here is an example of a poor item that was taken from an actual test for TV repairmen.

> 1. *The advantage of a full-wave rectifier over a half-wave rectifier is*
>   - a. *requires a center-tapped transformer.*
>   - b. *has a higher peak inverse voltage.*
>   - c. *has a higher average voltage for a given peak voltage.*
>   - d. *has a higher peak output for a given a.c. input.*

Another poor example follows:

> 2. *One disadvantage of the electrolytic capacitor is*
>   - a. *the electrolyte dries up quickly.*
>   - b. *the thickness of the dielectric makes it bulky.*
>   - c. *the high power loss.*
>   - d. *they cannot be used in a.c. circuits.*

The second item could be better stated this way:

> *What is one disadvantage of electrolyte capacitors?*
>   - a. *The electrolyte dries up quickly.*
>   - b. *The thickness of the dielectric makes it bulky.*
>   - c. *They have a high power loss.*
>   - d. *They cannot be used in a.c. circuits.*

Now each alternate follows the stem grammatically.

Probably the most common fault found in tests is the use of specific determiners. A specific determiner is an unwanted clue in either the stem or the alternatives that may tell the student a particular alternative is either correct or incorrect. The following paragraphs will describe some of the more common specific determiners.

Some words, such as *always* and *never*, tend to appear in false statements. Most students soon learn this, and use this clue in selecting an alternative. For example:

> *A paper capacitor is used*
>   - a. *always in high frequency, low voltage circuits.*
>   - b. *in low frequency, high voltage circuits.*
>   - c. *in low frequency, low voltage circuits.*
>   - d. *in place of larger air-type capacitors.*

The *always* in alternative "a" gives away an incorrect answer.

Another specific determiner is the length of the best alternatives. Often the longest alternative is the correct one because it completely covers all the facts of the question. Other times, the short, precise alternative stands out as the correct one. Whenever possible, all the alternatives should be of equal length. The test writer should take as many pains with the incorrect alternatives as he does with the correct one.

Sometimes a previously unused technical term is used in an alternative. The student is quick to realize he has not had the term as part of his training, so he will not select that alternative. On the other hand, including a highly technical term from the training materials in only one alternative may draw students to select that alternative.

Do not try to trick the students, and never use or misuse technical terms as clues in a test item.

Sometimes the test writer will repeat some of the terms from the stem in the correct answer. This is another form of specific determiner unless the same terms are included in the other alternatives.

This alternative illustrates how the use of a similar technical term reduces this item from a four-choice to a two-choice question.

As more capacitors are connected in parallel,
- *a.* *capacitive reactance will decrease.*
- *b.* *circuit current will decrease.*
- *c.* *applied voltage is divided proportionally.*
- *d.* *capacitive reactance will increase.*

Another type of specific determiner occurs when an alternative includes or overlaps another alternative. If the shortest alternative is correct, then the other alternative, by including the shorter version, is also correct. These overlapping alternatives can confuse the student since he is now confronted with several correct answers.

Just as bad as overlapping alternatives are opposite alternatives. It is very easy for the test writer to fall in the trap of writing at least one alternative that is just opposite of the correct alternative. In a way, this reduces the item to a true-false test. The student can usually pick the correct alternative from the true-false pair without really knowing the correct answer. If the opposite alternative is a logical alternative, it can be included in the item by writing the remaining two alternatives as another pair of opposites, neither of which is the correct answer.

The following is a good example of the opposite to the correct alternative being a logical alternative:

> *As the frequency increases,*
> - *a.* *capacitance decreases.*
> - *b.* *capacitive reactance increases.*
> - *c.* *capacitance increases.*
> - *d.* *capacitive reactance decreases.*

Unless they are properly used, "none of the above" and "all of the above" are specific determiners. If used at all in a test, these two alternatives must be the correct answer in about 25 percent of the items in which they appear. Thus, if "none of the above" is used four times in a test, it should be the correct answer only once. Otherwise, the student will soon realize that "none of the above" is probably the correct answer any time it appears. When testing the students' ability to make a fine discrimination among several alternatives, do not include "none of the above" as the last alternative. Both "none of the above" and "all of the above" are best used in items that test facts. Here is a poor example of a "none of the above" item.

> *A good photographic print has*
> - *a.* *good shadow detail.*
> - *b.* *a wide range of tones.*
> - *c.* *medium contrast.*
> - *d.* *none of the above.*

"None of the above" just does not fit grammatically with the stem of the item.

Here is a better example of this type of item.

> *What would be the general appearance of a film which had not been exposed but had been processed?*
> - *a.* *Varying shades of gray.*

b. *Black and opaque.*
c. *A negative.*
d. *None of the above.*

Testing Concepts

The following examples are one way to measure concepts using multiple-choice items.

1. *The symbol + means*
   a. *add.*
   b. *divide.*
   c. *multiply.*
   d. *subtract.*

2. *In woodworking, the symbol " means*
   a. *angle.*
   b. *center line.*
   c. *cut here.*
   d. *inch.*

Independent Items

It is important that each item in a test stand on its own. When items overlap, when one item supplies the answer to another item, the quality of both items is lowered. Here are two items that overlap. What is particularly bad is that these two items appeared one after the other on the test.

1. *The two main frequency bands which make up the frequency spectrum are*
   a. *power and radio.*
   b. *radio and television.*
   c. *d.c. and a.c.*
   d. *audio and radio.*

2. *The frequency which divides the audio and radio frequency bands is*
   a. *20 cps*
   b. *20 kc*
   c. *20 mc*
   d. *20 us*

This kind of overlapping can best be avoided by omitting one item rather than trying to alter one or both items.

Fine Discriminations

One type of item stands out when it is necessary to test the students' ability to discriminate among several factors. This item has no formal name, though it has been called a "nab" item. However, an example should be sufficient to show how one is constructed.

> *What portion of a contact print test strip would you examine to determine the correct exposure?*
> *a. Highlights.*
> *b. Shadows.*
> *c. Both a. and b.*
> *d. Neither a. nor b.*

This type of item proves to be a powerful tool for eliciting thought-provoking fine discriminations. It is also very useful when the concept or principle being tested offers only one logical wrong alternative. One word of caution: Alternatives c. and d. should be the correct answer a proportionate amount of the times this type of item is used.

Position of Correct Alternative

Beginning test writers tend to shy away from the first alternative being the correct answer. In fact, the third position tends to be the correct answer a disproportionate number of times. If the student gets the feeling that the first alternative will rarely be the correct answer or that the third alternative is more likely than not the correct answer, he will make his choices accordingly.

The goal should be to have the correct answer appear in each position an equal number of times. This can be accomplished several ways. The position of the correct answer should be selected on a random basis. No pattern of the correct answer position should appear in the final test. One way to insure random selection is to use a random-numbers table. Another way is to put four slips of paper in a hat, each marked with a letter from *a* to *d*. Pick out one slip and use the letter indicated as the correct answer position. Put the slip back in the hat, shuffle the slips and repeat the selection process for the next question. Be sure to put the slip back in the hat each time. This process gives each position an equal chance at being selected.

Multiple-Choice Check List

Some of the principles that should be followed in the construction of multiple-choice items are listed below. They can serve as a check list during writing, editing, or analyzing individual items or a whole test.

- Put everything that pertains to all the alternatives in the stem of the item. This helps to avoid repetitious alternatives and saves time.
- Word the stem so that it does not give away the correct answer.
- Avoid the use of synonyms for the correct answer in the stem.
- Keep all alternatives equal in length if at all possible.
- Design items which call for knowledge essential to the job. Avoid testing background knowledge or unimportant facts.
- Test the discriminations that the student has to make on the job.
- State each item in the working language of the student and the job. Unless called for by the job, do not test the students' ability to work with difficult language.
- Avoid the use of "a" or "an" at the end of the stem. They may give away the correct answer. Each alternative should fit grammatically with the stem.
- Make each item independent of every other item in the test. One item should not reveal the correct answer to another item.
- Use graphics and photographs when they can present a more job-like situation.
- When alternatives are numbers, they should be listed in ascending or descending order of magnitude.

- All alternatives should be plausible to the student who has not achieved the behavior required by the objective being tested.
- Catch questions, ambiguities, and leading questions should be avoided.
- If a negative term has to be used, emphasize the negative. Insure that students miss the item for the right reasons, not because they overlooked a negative term.
- Do not use double negatives in the item.
- Each item should test a concept or idea that is important for the student to know, understand, or be able to apply.
- Each item must be stated so that subject matter specialists would agree on the correct response.
- Edit each item to make it as brief and concise as possible.

APPENDIX D

CONSTRUCTING AND USING PERFORMANCE TESTS

A performance or skill test, as the term implies, is a test which requires a student to accomplish a job-like task under controlled conditions. This discussion will consider in more detail the two main elements of a performance test—job-like tasks and controlled conditions.

Performance tests emphasize the non-verbal; that is, they require the student to do something instead of just telling how to do it—"show and tell" rather than just "tell." Of course, all jobs require some use of verbal communication; for example, an automobile mechanic must be able to read repair manuals and parts lists. A performance test for this job could involve the student's ability to locate, read, understand, and follow certain technical materials. The fact that a certain task involves the use of words does not keep it from being a performance test as long as it is job-like and is derived from the training objectives. Under certain conditions, multiple-choice, completion, and short-answer test items can be performance test items; usually, however, they do not approximate activities that are job-like. Many times these kinds of test items are concerned with telling about a job instead of performing tasks required as part of a job. The aim should be to make a performance test as much like working on the job as is practicable.

Controlled conditions for skill testing means two things. It means that each student should be tested under conditions that will give him the best possible chance to display the skill which the test is to measure, and it also means that the test conditions should not change from one student to another. Keep these main points in mind as the development of performance tests is discussed.

Measuring Individual Proficiency

Being proficient at a given task is not an all-or-nothing matter. It is not likely that a student can either do something or not do it. What is more likely is that the students will vary as to how well they can do a given task. When measuring proficiency, the range of performance must be measured. Giving the test to some trained entry-level workers will determine a practical range of performance. If the training objectives state how well each required task must be done, few subjective decisions will have to be made. Once the level of proficiency has been specified for the objectives, and for their derivative performance test items, testing results are on an all-or-none basis. The student either meets or does not meet the minimum requirements.

Relation of Performance Tests to Written Tests

Many instructors speak of written tests and performance tests as if they were separate and distinct types. Written tests are often termed poor measures of proficiency, while performance tests are thought to constitute the only real measures of performance. This is not necessarily true. Some paper-and-pencil tests are indeed job-like and thus are performance tests. On the other hand, just because a test item involves equipment and requires the student to perform something does not mean that it is job-related. Also, the fact that performance is involved does not assure accurate measurement of ability. The real requirement is that the test situation make demands of the student that are as similar as possible to the demands of the task when it is done on the job. It is *what is measured* that counts in a performance test—not just the procedure by which it is measured.

Another thing to remember is that all tests that have the student perform are not necessarily good tests. The final test may not be a good measurement device because time considerations may force cutting the test; equipment considerations may limit testing; and the test procedure itself may make it impossible to test whether or not the student knows how to go about the task being tested. A performance test that has to be administered to one man at a time should measure something more than a test which can be given to a whole group at a time. Often a portion of what the performance test measures could be measured better by an objective written test. Be sure the performance test really measures "doing" and not just "knowledge about doing," which can be measured better in other ways.

How Performance Tests Aid Learning

Every test serves a function beyond measuring skill and knowledge. Tests provide specific goals for both the student and the instructor. There is nothing wrong in "teaching the test" if the test truly represents the training objectives. Of course, it is wrong to teach specific answers to given questions. If the instructor teaches the test and the test reflects the key terminal objectives, teaching the test is not an undesirable practice. Obviously, the better the students are prepared for the criterion test, the better they can meet objectives. Experience has shown that when skill tests are part of the training program, the training itself will tend to be altered in the direction of teaching what the test measures. This is desirable as long as:

- The content of the test is a good representation of all the training objectives.
- Teaching the test does not allow students to short-cut learning to the extent that they only learn that which is specific to a given test.

Performance tests also aid learning by providing feedback to the student. The tests give the individual student a means of identifying his areas of strength and weakness. He may be able to correct these deficiencies in later parts of his training or on the job. Thus the skill test can have an effect on student learning both before and after test administration.

Suggestions for Performance Test Construction

There is a very distinct difference in the rules for writing skill test items and those for knowledge test items. The items in a skill test are the tasks the student must perform or the decisions he must make when he solves an on-the-job problem or works at a job-like task. Thus, the item sequence is related, instead of independent, as in the knowledge test. The interdependence of skill test items means that an error early in the sequence may have a marked effect on the final outcome of the task.

For example, in a trouble shooting sequence, an early

incorrect decision that the malfunction is caused by a particular part within the truck's electrical system may make it impossible for the student to find the fault or at least delay his finding it beyond the acceptable time limits. In another example, the student takes certain actions based on certain information he has at the time. He is then committed to the effects of his actions, even though these actions were not the best. Connected sequences within skill tests mean that each student may take a somewhat different test because his test depends on what he has done up to that moment. This is very different from knowledge tests in which each item presents the same challenge to each student, whether the item occurs early or late in the test.

One problem in performance test construction is how to overcome this interdependence of sequenced responses. If the test is structured too rigidly to keep all students on the same track, the situation is unreal and unrepresentative of actual job requirements. If the test is completely realistic and allows each student to go his own way, the conditions are no longer controlled. Thus, there must be enough structuring to make each test administration somewhat comparable, while at the same time allowing each student leeway to perform as he would on the job. One way to combine realism with uniformity is to put the student back on the track (after making a note of the error) whenever he deviates too far from accepted performance. This is not really as artificial as it may seem at first glance, because in a real work situation his supervisor would probably do the very same thing. The instructor can play the role of the supervisior and still make the test realistic. He can simulate the natural job functions of the supervisor, including inspecting the work at key points; and, having found that the student has made a mistake, tell him how to correct the mistake, and then allow him to proceed with the remainder of the task or tasks. This kind of test construction will lead to the problems of weighting and scoring the test, which are covered later in this section.

Test Materials

The various tools and equipment of a trade are an important part of skill testing. The student usually works with most of the tools and equipment of the job. His job activities range from the operation and maintenance of equipment to making decisions based on information furnished by the equipment. Since skill tests attempt to duplicate the job situation, it is reasonable to assume that actual job materials and equipment should be used in the skill test. While this is essentially a sound practice, there are several factors that must be kept in mind:

- The actual equipment may exist in such a small quantity or be used so constantly for ongoing work and training that it is not feasible to use it in skill testing.
- The actual equipment was designed for a specific use which may not be compatible with the evaluation of student performance. The parts of the equipment not directly involved in the test may cause three sources of difficulty:
  (1) A malfunction may occur in part of the equipment not used in the test. This may make it necessary to discontinue the test until repairs have been made. Also, a malfunction may go undetected, and influence test results, without the test administrator or the student knowing of such influence.
  (2) The non-related parts of the equipment may interfere with the test tasks, and decrease the student's opportunity to display the required skills.
  (3) There is danger of damage to operational and sometimes expensive equipment.

Simulators

Because of the difficulty in using real equipment during skill testing, simulators can sometimes be used more effectively. Simulators can be specially constructed for skill testing, or they may already exist as part of the training equipment. It can be argued that simulators do not constitute a *really* job-like situation. Carefully designed simulators can be effective and yet concentrate on the parts of the job relevant to skill testing. There are four main arguments in favor of simulators:

- The simulator can isolate for measurement that

portion of the job activity directly involved in the assigned task.

- The simulator can be relatively inexpensive in comparison to actual equipment. This makes it feasible to have several simulator units for testing several students at once.
- The simulator can usually be constructed to insure trouble-free operation and to assure student safety.
- The simulator can be constructed to allow a series of pre-set malfunctions to be induced by merely throwing a switch. (It is often highly impractical to induce malfunctions in the actual equipment.) An added advantage of a simulator with built-in malfunctions is they can be "repaired" just as simply—just flip the switch.

Depending on the purpose for which they will be used, simulators can vary all the way from simple pictorial representations of job situations to pieces of equipment that practically duplicate the actual equipment. The main objective is to design a simulator that is appropriate to the measurement of the skills being evaluated by the test. Here are some guidelines to insure that this objective is met:

- Eliminate those parts of the equipment which interfere with the performance of the skill which is being measured. (If actual equipment is being used, it is a good idea to remove protective plates, housings, and dust covers. The time the student would spend to remove these items may not contribute significantly to skill measurement.)
- Eliminate those parts of the equipment which are not part of the actual test. This will usually decrease the space required for testing. It will also allow the student to perform without wasting time or energy. If it will add to test realism, the removed parts can be represented by symbols, block diagrams or other illustrations.
- Reduce the complexity of the equipment to reduce the chance of malfunction in those parts of the

equipment not contributing to skill measurement. Many times complex electronic circuits can be reduced to simple circuits that will provide the same reading on the test equipment as would the actual equipment.

- Build a model representation of the actual equipment. These models should allow the student to respond in such a way as to indicate how he would perform on the real equipment. This technique is often used to demonstrate knowledge of a procedural sequence. Care should be taken to insure that such a technique not only measures job knowledge but job skills as well.
- Substitute a verbal description of acts for real acts. For example, a procedure sometimes used in testing trouble shooting consists of presenting the student with a list of possible checks or tests from which he chooses the check he wants to make. He is then informed of the hypothetical result of this check or test and proceeds to the next cycle of check and feedback. The student identifies the source of trouble when he thinks he has found it. He is then told if he is right or wrong. All during the test, a record is kept of the student's trouble shooting behavior. Again, it is important to be sure that the test measures job skills and not just job knowledge. Use this technique only when it is not feasible to measure actual trouble shooting during skill test situations.

Use of Graphics in Performance Tests

Many on-the-job tasks are based on equipment or situations which can be communicated more easily by graphic illustrations than by words. The following methods are available for using graphic items:

- A verbal problem with alternatives shown on a graphic illustration.
- A verbal problem with verbal alternatives based on an illustration.
- A verbal problem with illustrated alternatives.

If you decide to use graphics as part of a skill test, the following rules will assist you in constructing a better test:

- The illustration must be necessary for the correct answer.
- The illustration should be clear enough to convey all necessary information.
- All nonessential information should be eliminated.
- Any information which might give away the correct answer should be eliminated.
- Reference to the illustration must be specific. The various parts of the illustration should be lettered. The letters should be arranged in a definite order to eliminate unnecessary hunting.

Technical Review

After the skill test has been constructed, the answer sheet designed, the graphics and equipment prepared, and the scoring and weighting decisions made, the items should have a two-part technical review. The first review should be by an expert in test construction. This review is used to help insure that an effective test has been designed. The second review should be by a subject matter specialist. This review insures that the test is technically correct. He should be concerned with both the problems and the alternatives, if any are given. He has to insure that the correct alternative is indeed the only unambiguous, correct answer.

Initial Test Tryout

After the technical review and rework, the next step is to try out the test on four classes of people if possible: experts with a great deal of experience on the job; working apprentices who have six months to a year of experience on the job; students who have just completed the training program; and entering students who are typical of those who will take the training program. There are several reasons for test tryout. One is to be sure that you have provided clear instructions for test administration. Another reason is to be sure the test items are job-like and can, in fact, be accomplished by experienced workers. You will also need to collect data which will be used to analyze the items, to determine the weighting, and to set the time limits (Details of test validation are in Chapter X.) During tryout, all the test instructions and conditions should be exactly as they will be during operational use of the test.

Test Administration

It is very important that both the test administrator and the student know exactly what they are to do. Perhaps the only way to insure this is to write a clear set of instructions for the test administrator and another set of instructions for the student. Normally, the instructions for the student are included in the test administrator's instruction booklet. Then, during the actual test situation, the test administrator reads aloud the instructions to the student, while the student is following in his own booklet. The test administrator's booklet should also specify all the tools, equipment, and other auxiliary materials required during the test. The best way to present this information is with a check list. Be sure to include the very obvious, but often overlooked, items such as paper, pencils, wire, solder, or any other items required during the test. Finally, the test administration booklet should prescribe the test conditions and time limits for the various parts of the test. At test time, each man should have a clear, well-lighted place to work that is as free as possible from noise and other distractions.

The purpose of a performance test is to measure how much knowledge of a given subject matter a student can apply. Experience has shown that, for skill tests, the rate of answering is little related to knowledge. Therefore, the student should be given enough time to attempt all parts of the test. There are some, of course, who never would have enough time; so some limit must be set. After the initial tryouts of the test, it is a good rule to set the time limit at the point where 85 percent of the students can complete the test.

Scoring Performance Tests

Since the student does not just answer questions in a skill test, there is a problem in determining how he should be scored. It is possible to evaluate both the process and the product of the process in skill tests. Some objectives are more concerned with how a student does something, while other objectives are concerned with the results or product produced without regard for the way in which they are produced.

Here are five conditions that support scoring the process:

- The steps in the procedure can be specified and have been explicitly taught.
- The correctness of the outcome or the quality of the product depends directly on the performance of a highly specified procedure.
- The extent to which an individual deviates from accepted procedure can be accurately and objectively measured.
- Almost all the evidence needed to evaluate performance can be found in the way the performance is carried out.
- Enough staff are available to observe, record, and score the procedures used during the skill test.

The following four conditions make it more appropriate to measure the product.

- The product of performance can be measured accurately and objectively.
- Much or all of the evidence needed to evaluate performance is found in the end product and little or none of the evidence is found in the way the performance was carried out.
- The proper sequence of steps to be followed in attaining a goal is indeterminate, or has not been taught in training. Although everyone knows the steps, they are hard to perform and skill is ascertainable only in the final product.
- The evaluation of skill test procedures is not practicable because staff are not available to observe, record, and score these procedures.

The actual techniques of scoring skill tests follow from the discussion above. With products, the skill can be scored according to the degree to which it meets the objective for the task. There are three principal kinds of scores: errors, accuracy, and speed.

The most common way to judge a skill is in terms of errors

made. Errors can be incorrect responses, choosing the wrong tool, selecting inappropriate materials, using the wrong procedure, and failing to observe a safety precaution. Omitting a response, making a wrong response, and making a response in the wrong sequence are also errors.

Accuracy refers to the amount of deviation from some standard. Actual measurement of the deviation is usually required. Thus, accuracy is measured when the final product is checked to determine whether it is within tolerances.

Speed is measured by determining how long it takes the student to accomplish each unit of task activity. Speed may also be the number of units produced in a given length of time.

Which measure or combination of measures should be used for a given skill test depends on training objectives. If accuracy is more important than speed in a given task, the test should be weighted and scored accordingly.

As mentioned earlier, many skill tests will require the presence of a "supervisor" to put the student back on the track when he goes too far astray. Because this outside help is usually given for those mistakes that would affect all the other tasks in the test, a negative score should be assigned for each assist. Thus, the student who required several assists would receive a much lower score than the student who required no assistance.

Administering Performance Tests

As with knowledge tests, two kinds of instructions are needed when administering skill tests—instructions for the student and instructions for the test administrator. Earlier, it was said that while the items on a knowledge test should be independent, the items on a skill test may occur in a sequence leading to particular results. Thus, the student's instructions must direct his attention to each item. He cannot afford to overlook any item. On the other hand, because the test should simulate the job situation, the sequence of items should be left up to the student. He should know the correct sequence. Thus, the skill test instructions should tell him what he is to do, but not how to do it. Random activity is not desirable, but the student should not be given a recipe either. The situation can be summed up with four rules:

- Be sure the directions are sufficiently clear so the student understands exactly what he is to do.

- Do not take the performance out of the skill test by furnishing too much cookbook information.
- Do not use a skill testing situation to test that which could be tested more efficiently with a paper-and-pencil test.
- Tell the student how his skill is to be rated. He should know whether his goal is to work for speed, accuracy, avoidance of errors, or a combination of these factors.

Performance Test Administration Problems

Once a skill test has been developed, its administration poses many problems that must be solved if the effort expended in making the test valid, reliable, and objective is not to be wasted. These problems center around the fact that the conditions during testing should be the same for all individuals tested. If the test conditions vary from man to man, their test scores will not be comparable.

To help insure that the test situation remains the same for each individual tested, attention should be given to the following factors:

- All equipment and tools used during the test should be checked for proper operating condition before each testing session.
- Arrangements should be made to insure that all required supporting equipment, such as hand tools and spare parts, is given to each student being tested.
- Test procedures to be followed by both the examiner and the student should be developed systematically. Written instructions should be provided for the examiner.
- All examiners should be trained in administering the test before actual testing begins.
- Adequate protection against compromise must be achieved. The ingenuity of students must not be underestimated.

Performance Test Construction Check List

Listed below are a number of questions intended to assist the test writer in developing, editing, and evaluating a skill test. These questions should call attention to the many decisions that must be made during test construction. Keep in mind the training objectives as each question is answered and the appropriate decision made.

- Have you stated as accurately as you can what you want the test to measure?
- Is the skill being measured representative of that indicated in the training objective?
- Do you want to use actual equipment or would some modification, or even a simulator, be better?
- Will equipment being used in the test permit the student to display the skill being measured?
- Will the tools furnished the student give him a good opportunity to exhibit the required skill?
- Are you primarily interested in evaluating the result of the student's effort (the product) or in how the product was achieved (the learning process)?
- Considering the skill being measured, what evaluation factor (speed, accuracy, or errors) should be used?
- If the test yields more than one measure of performance, how can these subscores be combined and weighted?
- If the test yields only a total score, is this score meaningful, or is it composed of inconsistent parts?
- Is the total score compatible with other grading systems being used?
- Will the skill test measure different abilities than could be measured by a written knowledge test?
- Are there minor activities which can be omitted from the test or from the scoring?
- Do the directions make it clear to the student exactly what he should do?

- Will each student tested face the same initial situation?
- Are the demands and conditions of the test as job-like as possible?
- Is the student told only what he should do, or is he told how he should do it?
- Are all the aspects of recording and scoring performance as objective as they can be made?
- Do the directions tell the student how he will be rated (accuracy, speed, errors)?
- Can a technical specialist complete the test satisfactorily?
- Has the test been tried out to determine the range of scores possible and to determine the minimum passing score?

Sample Performance Tests

This section will present some acceptable examples from a variety of skill tests. While these are only excerpts, a complete skill test should have instructions to the test administrator, instructions for the student, a list of tools and test equipment for use by the trainee, a spare parts list, and a list of tools, bad components, and special equipment needed by the test administrator. In addition, it should contain the test items, procedures, and a scoring key.

Performance Test 1

Instructions
Television Repairman Test

"You will find on your bench an XYZ television set. You will also find various tools, a power supply, and pieces of test equipment.

"You are to perform an operational check on the TV set. If it does not operate properly, you are to trouble shoot the set using any of the tools and test equipment you choose. When you localize any malfunction, perform the necessary repair operations to put the set in operating order."

(The student is told only the information given above. The

test may involve, depending on the malfunctions induced in the equipment by the test administrator, a wide variety of tasks and subtasks.)

Performance Test 2

Instructions
Service Station Mechanic

"Today you will be working on problems that have actually occurred as a result of service calls. Since these are on-the-road problems, we want you to use field solutions; that is, use any technique or method you can think of. You will be scored primarily on your ability to get the vehicle back in operation. You won't be scored on the method that you use to do this.

You will be allowed up to three hours to complete the test. You will be given a series of tasks in the areas of trouble shooting, on-site adjustment, operation, and practical maintenance.

There is a time limit for each problem. You will not know how much time is allotted, but you should work as rapidly as convenient. Don't be discouraged if you don't finish a problem before I call time. The test is designed so that no one is likely to finish all the items. When I call time, stop working so we can go on to the next problem."

Performance Test 3

Instructor's Check List
Use of the Tube Tester

Checkpoint	Score one point if repairman performs as indicated.
1. Finds tube in Tube Data Booklet.	....
2. Sets switches before plugging in tube.	....
3. Sets switches correctly.	....
4. Tests for shorts first (both sides).	....
5. Corrects condition of tubes.	....

6. Leaves equipment and area in safe condition. ....

Performance Test 4

Examiner's Check List
Pre-start Check on Truck

Check point	Score one point if driver performs as indicated.
.....1. Lights	Must make visual check of both high and low beams.
.....2.	Performs action indicated.
.....3.	Visual check.
.....4.	Uses pressure gauge.
.....5. Tank	Fills Tank.
.....6.	Visual check.

Performance Test 5

Instructions to Student

1. Using the tools, parts, and equipment on the test bench, construct the bridge circuit shown below:

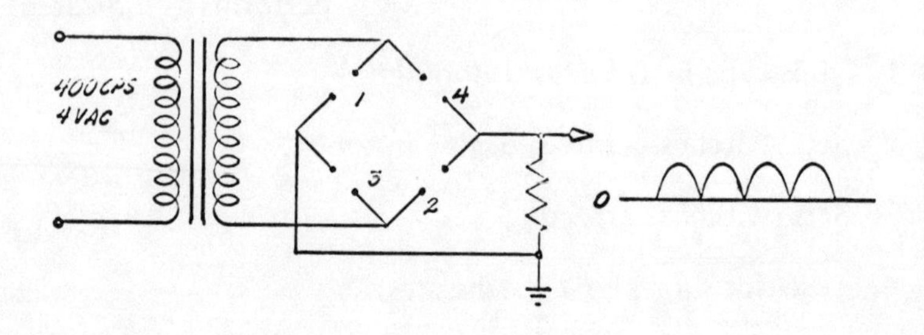

2. Have your instructor check the circuit. Instructor check:________
3. Adjust the test bench audio oscillator to a 400-c.p.s. sine wave output with an amplitude of 4 volts r.m.s. Use the appropriate piece of test equipment to make bench adjustment.
4. Connect audio oscillator output to the primary of T_1.
5. Using the oscilloscope, draw the waveshape impressed across the T_1 secondary.
6. Calculate the peak-to-peak value of the waveshape:______
7. Place probe across R_1 and record the output waveshape:______
8. Remove diode number 2.
   Draw the output waveshape.

Performance Test 6

Training Objectives: Use the diode tester, oscilloscope, and trace patterns to determine the condition of unknown diodes, in terms of good, fair, or bad.

Skill Test: Using plug-in leads with alligator clips on the end and the diode assortment, place each diode across the diode terminals of the diode tester. Compare the oscilloscope display with the diagrams below to determine the condition of each diode.

Diode assortment number is ______ .

Diode	*Condition*
1	____________
2	____________
3	____________
4	____________
5	____________
6	____________
7	____________

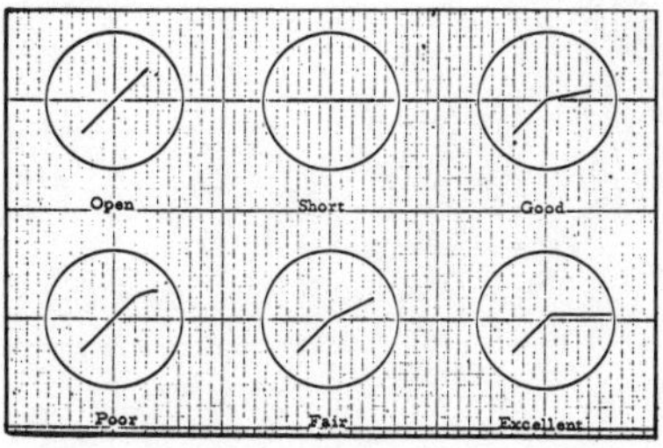

Have instructor check your findings.

APPENDIX E

PREPARING AND USING RATING SCALES

Ratings have been and will probably continue to be one of the most widely used evaluation devices. A rating is human judgment which has been converted to some kind of a scale (for example: superior, average, below average). Ratings are used to measure human or product characteristics for which other tests are not available. Ratings are also used as the standard by which the validity of other measures are judged. It has been often said that a student is as good as his teachers say he is; other evaluation devices can only approximate this type of judgment.

At any school you can probably hear statements something like the following: "Wilkins will make a good carpenter." When questioned as to how he knows Wilkins will make a good carpenter, the instructor will reply, "Because I just know, that's all." This situation is indicative of the main shortcoming of ratings: the halo effect. Raters tend to judge people in terms of an overall impression of their worth. This overall impression does not take into consideration the lack of relationship between certain characteristics of a student and his ability to do a job.

As an example, suppose a certain school is conducting a highly successful course for TV repairmen. The course objectives have been selected from significant job requirements. Over 85 percent of the students have mastered all the objectives when they graduate, and they are very competent in their ability to perform job-related skills. The students gain employment in a variety of TV shops in different cities, and after a few months, their new supervisors are asked to rate them and their repair skills.

These former students are all highly competent, but they differ in personality, motivation, discipline, initiative, dependability, and neatness. Because of the halo effect mentioned earlier, the judgments of their job proficiency will accordingly reflect the supervisors' reaction to these other characteristics. Ratings also suffer from the "error of leniency." Most people prefer not to say bad things about others, so a student is likely to be rated above average in socially desirable traits.

Single ratings are not very reliable, nor are they very objective. The personalities of both the student and his rater interact to produce a rating which often does not reflect the true performance of the student. The tendency of ratings to lack objectivity can sometimes be overcome by having the student rated by several supervisors. However, in most training situations, it is difficult to find several staff people who know an individual student and his job performance well enough to rate him.

Ratings are hard to standardize. Each rater has his own standards and frame of reference which he brings to the evaluation situation. There are "hard raters" and "easy raters," and there is no manageable way to compare ratings made at different times and places on different students.

For all these reasons, ratings may be the poorest technique for evaluating the ability of students to achieve training objectives. Nevertheless, there are times when there is just no other way to assess behavior skill, especially in working and living habits and attitudes. Moreover, skill performance tests often require the use of rating scales because even though the tasks may be carried out correctly, some students will carry them out better than others. Ratings can be valid if they measure the trait or factor they are supposed to measure, and ratings can be reliable if they measure traits or factors consistently. This Appendix is concerned with helping the course designer to prepare the best possible rating scales.

Rating Methods

There are two general classes of ratings: relative and absolute. Relative rating is often used to rate several persons or a whole class, with each student being evaluated only in relation to his peers. This is a very undesirable approach to evaluation when the real goal is to determine how well each student has achieved the training objectives.

Thus, rank order and paired comparison ratings should not be used. Rather, one of the several absolute rating methods should be used in vocational training programs. In the absolute method, the instructor or supervisor must assign an absolute value to the trait, performance, or ability of an individual student without reference to the performance of any other student. This can best be accomplished when the observer rates according to a fixed scale. The rater depends on his past experience and training to assist him in rating the student.

The remainder of this section will discuss various absolute rating methods.

Numerical Scales

Probably the simplest form of an absolute rating is the numerical scale. While any number of points is feasible, an odd number is often used so the center figure can stand for average or mid-range. Although the term average is used here, remember that it is not based on the average student, but rather it is *based on the performance of the average entry-level worker.* Thus, a seemingly relative scale is actually absolute for rating students. The number of points on the scale should depend on the number of differentiations required, and, even more important, on the ability of the rater to differentiate. A three-point scale may provide all the differentiation required. Most raters can make five differentiations. It takes a highly trained rater to make as many as nine differentiations reliable. For these reasons, practical rating scales should only contain three to seven points. One example of a five-point scale would be: five, highest; four, above average; three, average; two, below average; and one, lowest. Another rating might call for points indicated by superior, average, and below average. Here is an example of a seven-point numerical scale used to measure the degree of acceptability of the student's work.

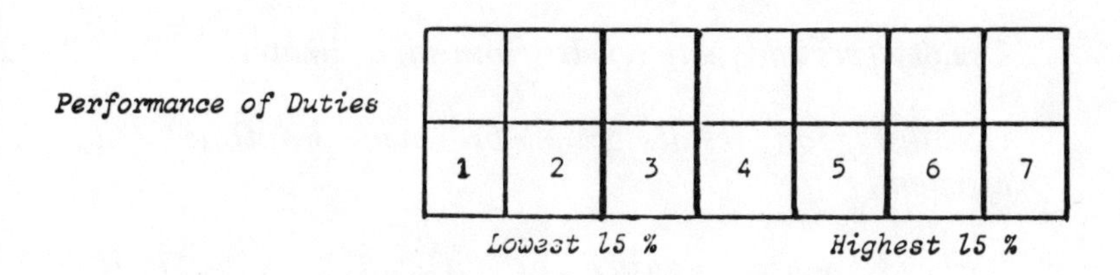

Descriptive Scales

The descriptive scale uses words or phrases to indicate levels of ability or performance. For example, here is a descriptive scale for rating metalworking ability.

Instructions: Place a check mark in the scale above the word that most accurately describes the worker being rated.

Poor	Fair	Good	Excellent	Outstanding

The major disadvantage in the use of descriptive words is a semantic one. An "excellent" metalworker does not mean the same thing to all raters. Another disadvantage is the difficulty in selecting words to describe degrees of performance which are evenly spaced. Many raters, for instance, feel there is less distance between excellent and superior than between fair and good.

An even better descriptive scale contains short phrases to assist the rater in making his choice. As an illustration, here is an example of a descriptive scale used to rate the ability of a person to supervise.

A poor supervisor. Little or no cooperation from subordinates.

Usually gets adequate results from subordinates.

Obtains good results from his men. Controls unit efficiently.

Succeeds under unusual or difficult circumstances. Secures high production.

> *Outstanding ability to get the maximum out of his men and all available resources.*

Descriptive scales are more versatile than numerical scales because raters have the tendency of always relating the lowest numbers with unacceptable performance. By using a descriptive scale it is possible to indicate satisfactory performance as the lowest possible rating. The remaining points on the scale would then be used to indicate degrees of more than satisfactory performance. A good choice of descriptive phrases helps to reduce the possibility of all students being rated average or close to it.

Check Lists

The check list is actually a two-point rating scale. Check lists are useful for rating a student's ability to perform a set procedure. They are also the best method of rating skills when the goal is to determine whether students have reached the minimum level of performance required by the objectives. Check lists are often based on yes-no, go– no-go, or satisfactory-unsatisfactory scales. Check lists tend to be reliable because they call for only two choices. Fewer choices require fewer and less refined judgments. If the choices on a check list are sound, the chances for both error and bias are greatly reduced.

Remember one key point. A check list can tell you yes or no, but not how well a task was performed. A metalworking check list will indicate whether the student followed the correct sequence for setting up the lathe, but it will not indicate how well the student performed the task above the minimum acceptable level.

However, there are techniques for making check lists powerful instruments for assessing student performance. Criterion-referenced check lists that do specify behavior and product quality that is critical to successful performance of the task are highly reliable means for judging competency, both in training and on the job. Because the development of such criterion-referenced check lists is so important to the evaluation portion of an instructional system, a separate appendix is devoted to the techniques. Turn to Appendix F for further details.

Product Scale

Many times the product of performance is rated rather than

the performance itself. However, there are problems associated with this practice. For example, an apprentice automobile mechanic is given a carburetor to repair; however, because he is still learning this task, he spends two days repairing the carburetor. When he is finished, the carburetor is in perfect operating condition. The end product is excellent, but the performance is poor because he took two days for a job that should take only two hours.

There is a danger in basing all evaluation on the finished product. Many times the student will eventually produce a finished product of high quality, but in the process of doing so, he might have done any or all of the following:

- Consumed too much time.
- Wasted raw materials.
- Asked for and obtained much more help than he should.
- Performed inaccurate or faulty work which is concealed in the finished product.
- Abused tools and equipment.
- Violated safety rules.
- Failed to follow correct procedures.
- Failed to learn the related information he was expected to learn.

If these points are kept in mind, rating the product has its place in training evaluation. Products, unlike performance, are usually tangible. Thus, product ratings are more reliable than performance ratings. When carefully constructed and used, product ratings can eliminate most, if not all, rating errors. The scale provides the rater with a tangible standard with which to evaluate the product. The halo effect should not be present because the rater does not even have to know who produced the product. Using the kind of check list device described in the previous section, in conjunction with product rating, is probably the most efficient and effective means of measuring job-like performance. In fact, criterion-referenced check lists should, whenever possible, include end-product rating items.

Improving Ratings

The best way to improve ratings is through training the raters. Every rater should understand common rating errors, correct rating procedures, and various rating methods. Next in importance is experience. Satisfactory reliability can be achieved through training plus actual experience as a rater.

The following guidelines can help increase reliability of ratings.

- Be sure that all raters understand how to use each scale. Make sure the points on the scale have a common meaning to all raters.
- Current observation of the task is more accurate than memory. Do not allow a long time lapse between observation of the task and the assignment of the rating.
- Do not allow the rating of one factor, trait, or step to influence the other ratings.
- Use all the points of the rating scales when they are deserved.
- Rate only specific, observable items of behavior.
- Allow enough time for observation before rating.
- Rating one-time behavior can lead to false ratings.
- If possible, use more than one competent rater.
- Combine their ratings for an average rating to increase reliability.

Check List for the Rating Device

Validation of each rating scale is a continuing process. The following check list may prove useful for constructing, analyzing, and improving rating devices and their use.

Construction

- Are the objectives of the training clearly defined?
- Are the factors or traits selected essential in the performance of the job?
- Are there enough factors to describe the performance without overburdening the rater?

- Is each factor defined in terms of observable behavior?
- Is the number of performance levels related to the training objective?
- Is the number of levels on the scale within the discriminative ability of potential raters?
- Is each level of performance described in terms of observable behavior?
- Have the initial weights for each factor been determined by pooling the judgments of several technical specialists?

Using the Rating Scale

- Are the ratings made as soon as possible after observation?
- Is the final rating based on ratings by two or more observers?
- Are the observers rating only those traits which they have had the opportunity to observe?
- Are all performance levels, from lowest to highest, used if they are appropriate?
- Is the rating scale understood by all raters?
- Are the ratings on one factor free from influence by other factors or an overall impression?
- Have the raters been trained in the use of the rating device?

APPENDIX F

CONSTRUCTING AND USING CRITERION-REFERENCED CHECK LISTS

A detailed breakdown of behavioral objectives and performance tasks in the form of a criterion-referenced check list should be used as the primary instrument for evaluating student job capabilities. Criterion check lists provide a practical and functional means for structuring the observations required for performance testing. The characteristics of a criterion check list are:

- list of events, activities, and actions that make up the task
- critical to task, and thus distinguish between effective and ineffective performance
- time and job sequenced
- standards derived from performance objectives
- clearly observable—the events either occur or they do not
- instructor needs only to note presence or absence of critical action
- the series of discrete observations provides a valid, reliable basis for judging overall proficiency

Performance testing based on criterion check lists, along with evaluation of the products associated with the tasks, serve several important learning and evaluation functions simultaneously. Criterion-referenced check lists can serve as:

- diagnostic and prescriptive pre-tests
- performance tests for evaluating and recording student job competencies
- a set of standards from which the student can determine

what and how well he has to perform

- guides for structuring and organizing learning activities that help the student determine and analyze his areas in need of more work
- quick review of lesson content
- the basic source of data for an Occupational Readiness Record which documents evidence of a student's experience and capabilities
- the basic developmental instrument for evaluating, validating, and modifying the instructional program

Regardless of the format of the learning materials, criterion-referenced check lists provide a major source of corrective feedback for student and program evaluation that is so vital to the concept of instructional system development. The two sample performance checks that follow should serve to illustrate the basic simplicity and utility of the criterion-referenced check list.

POWER MECHANICS

Family: Auto Mechanics & Related Occupations.

Exit Level: Service Station Attendant & Related Occupations. (915.867)

LEVEL I

REMOVE AND REPLACE THERMOSTATS

PE 10-8

TASK	C.O.
21	5

NAME ____________

DATE ____________

MAN-HOURS ____________

INSTRUCTOR ____________

CRITERION CHECKLIST

L M S

☐☐☐ Removes and replaces thermostats.

U S

☐ ☐ 1. Correctly answers 80% (5) of the training check questions.

☐ ☐ 2. Drains coolant in clean bucket (if to be reused), checks for contamination, performs visual inspection of system, secures drain cock and removes hose if necessary.

☐ ☐ 3. Locates thermostat housing and removes appropriate bolts.

☐ ☐ 4. Removes and saves retainer ring (if present) and removes and inspects thermostat.

☐ ☐ 5. Cleans gaskets surfaces to a like-new condition without scrapings or foreign material entering system.

☐ ☐ 6. Replaces thermostat and gasket in proper position, with copper "pill" end in water jacket. End marked with "TOP" or similar markings are toward radiator.

☐ ☐ 7. Replaces housing and tightens bolts without damage or undue stress to housing, bolts, and threads.

☐ ☐ 8. Refills cooling system, tests for leaks, and readjusts coolant level following correct procedures as identified in the behaviors for Units on Visual Inspection and Filling Cooling Systems.

☐ ☐ 9. Performs tasks in an appropriate amount of time.

QPS/AIR/ABLE GENERAL WOODWORKING LEVEL I	RAIL AND APRON CONSTRUCTION PE 6-9	NAME______
	TASK # 53 / C.O. # 1 / T.O. #1-4	DATE______ MAN-HOURS______ INSTRUCTOR

CRITERION CHECKLIST

L M S

☐☐☐ Identify common adhesives by their appearance and application.

U S

☐ ☐ 1. Selects a resourcinol glue and states its function and application.

☐ ☐ 2. Selects a casein or resin glue and states its function and application.

☐ ☐ 3. Selects a polyvinyl acetate (white glue) and states its function and application.

☐ ☐ 4. Selects a contact cement and states its function and application.

L M S

☐☐☐ Select adhesive application tools and state how they are used and maintained.

U S

☐ ☐ 1. States common methods for applying glue.

☐ ☐ 2. Selects proper solvent for the four common glues.

L M S

☐☐☐ Select the appropriate adhesive and tools and cement members of materials together so that a hairline joint exists between members.

U S

☐ ☐ 1. Uses appropriate glue and glue applicator for cementing veneer.

☐ ☐ 2. Glues members of material so that a hairline joint exists between glue members.

Given various kinds and sizes of screws, boring bits, a blueprint and stock:

L M S

Identify screws by their shape and size.

U S

1. States the correct size and shape of wood screws using a wire gauge.

L M S

Drill a pilot hole to the root diameter of a screw.

U S

1. Uses correct bit for drilling a pilot hole required by the root diameter of a screw.

L M S

Insert a soft wood plug to increase screw holding power in end grain.

U S

1. Drills and inserts soft wood plugs (with glue on them) to specifications.

L M S

Counterbore a screwhole for inserting a plug to cover a screwhead.

U S

1. Selects proper bit and size to match a plug required.

2. Drills counterbore to correct size and depth specified.

3. Cuts plug to size having the grain going in the direction of that on the face stock.

4. Makes the part to within 1/32" of blueprint dimension requirements.

APPENDIX G

SELF-SCORING RESPONSE AND FEEDBACK DEVICES

Throughout previous discussions of the systems development process and product, the need for constant and immediate feedback, to both the student and the instructor, from the frequent tests has been emphasized. It should be apparent that unless the student is given the responsibility for handling a majority of the test-scoring and recording chores, the instructors will be overwhelmed with mere clerical tasks, leaving them no time for their primary duties as learning managers and tutors. The instructor should not be burdened with correcting pre- and post-tests of prerequisite skills and knowledge called for by the interim objectives. Moreover, the instructor must be free to handle the more critical task of overseeing and conducting the performance checks and tests demanded by the terminal objectives.

In addition, the students have an immediate need for the feedback that is a part of the learning process itself. If the students must wait for the instructor to correct all the internal pre- and post-tests and then to provide corrective feedback, the entire individualization effort would soon break down completely. The entire instructional systems concept is dependent on the immediacy of the differential corrective feedback. Thus, there is a very real need for testing devices that are self-correcting, self-scoring, and self-diagnostic and which provide immediate knowledge-of-results based on the disclosure of wrong responses and the confirmation of correct responses. Such devices do exist, but the range in complexity and cost is considerable. The range includes the following techniques and devices:

- specially prepared answer sheets that reveal the correct response when marked with special marker pens.

- specially prepared answer sheets that require erasing to reveal the answer under an opaque overprinting
- punchboard testing devices with which a stylus is used to punch out answer choices
- mark-senses on IBM-type answer sheets that can be scanned and scored electronically
- electro-mechanical testing devices (teaching machines) which indicate right and wrong responses
- computerized testing programs
- computer-managed instruction
- computer-aided instruction

All of the preceding devices and techniques display certain common characteristics that are essential to the performance data feedback requirements of an instructional system. In varying degrees, each shares the following capabilities:

- provide for at least four or more multiple-choice responses
- provide immediate knowledge of wrong responses
- provide immediate revelation of the correct response
- provide a permanent record, item-by-item and choice-by-choice, of the students' response patterns
- provide item analysis data

Using such devices promotes individualization in the following ways:

- relieves the instructor of a tremendous amount of clerical work which is essential to the operation of an instructional system but which he cannot possibly handle by himself
- puts the responsibility for making meaningful diagnostic and prescriptive learning decisions with the student, where it belongs, but does maintain the integrity of the testing situation because the devices are relatively "cheat proof"
- provides immediate knowledge-of-results that is integral to the learning process
- provides a permanent record of student (and system)

VAN VALKENBURGH, NOOGER & NEVILLE, INC.

NAME C. Foltz RIGHT 6 WRONG 4

CLASS/COURSE Woods MODULE/TEST NO. 4-2 SCORE 23

DATE 3/11/70 TIME 9:00-9:20 ITEMS OF DIFFICULTY 4-6-8-10

Self-Scoring TRAINER-TESTER* Response Card

PATENTED • VVN&N • TRADEMARKED

Directions—Variable Alphabetical Response Mode:
Erase the block where you think correct answer is. Preferably use clean, firm, non-plastic pencil eraser, with reasonably sharp edge. Your instructor will **designate** the **correct** answer response for a **particular exercise**, for example:

Correct Answer Designated:		**Then other responses are:**	
"T" = Right	etc.	"H", "E" or "L" = Wrong	etc.
"H" = Right		"T", "E" or "L" = Wrong	

If your instructor wishes you to learn the correct answer, continue erasing until the response designated as correct is revealed; make as few erasures as possible. For self-scoring, grading and item-of-difficulty identification see Direction Sheet.

Item of Difficulty Mark

ITEM NUMBER (UNIT)	(a)	(b)	(c)	(d)	SCORING POINTS
1	T				3
2				T	3
3		T			3
(4)			L	T	2
5	T				3
(6)		L	T	H	1
7			T		3
(8)	T	H	E		1
9	T				3
(10)		E	L	T	1
11					
12					
13					
14					
15					
16					
17					
18					
19					
20					23

ITEM NUMBER (UNIT)	(a)	(b)	(c)	(d)	SCORING POINTS
21					
22					
23					
24					
25					
26					
27					
28					
29					
30					
31					
32					
33					
34					
35					
36					
37					
38					
39					
40					

T-T No. Z-11 Second Series

APR. 1969

performance that can be analyzed in detail later

- provides the data base for assessing and adjusting system performance, both immediate corrective actions and long-term revision

Most of the simple paper and pencil, self-scoring devices do serve adequately the needs of a classroom feedback system; yet are considerably less expensive and simpler to use and to maintain than electro-mechanical devices and computers. Once again, the least expensive medium that will produce the desired results is the logical choice. Although there are several comparable self-scoring answer sheets on the market, all the sample self-study learning units that are contained in the next three appendices have been keyed to the "Trainer-Tester" device illustrated on the preceding page. This erasable self-testing device has proven, in most situations, capable of producing the data for corrective feedback that is so essential to the instructional system.

APPENDIX H

SAMPLE SELF-PACED WORKSHOP MATERIAL

A Basic Course in Electronics Circuitry Repair

The following is a sample lesson of a self-instructional laboratory course in basic electronics circuit maintenance and repair. The course is based entirely on "learning-by-doing" workbench projects. There are no lectures on electronics theory, only self-study, "hands-on" workbench exercises occasionally supplemented by study in a programmed text on electronics fundamentals.

Notice the high level of student activity and involvement. The student must continually perform and respond to questions, and then check his performance and responses. Notice particularly the three different levels of assessment and confirmation. Written responses are required throughout the workbook. The many interim steps have the correct answers supplied "in the open"—the student can confirm his response by simply reading the next sentence below the blank to be filled in. On the other hand, terminal behaviors are checked *by the "cheat-proof" erasing answer key for self-scoring.*

Finally, the student must demonstrate his attainment of the terminal objectives by actually performing the key tasks for the instructor. Of course, there are also end-of-the-unit and end-of-the-course criterion tests. These criterion tests are also performance tests that require the student to demonstrate his proficiency for the instructor. Some of the criterion test items from this course are included in the following pages as examples of the performance test.

Lesson

Direct Current (d.c.)
Circuits

Introduction

This project has three parts. The first portion covers series circuits, the second portion covers parallel circuits, and the third portion is on series-parallel circuits. The information in this project marks the real beginning of your electronics education. You will be using combinations of these circuits throughout this course and throughout your career. Study the circuits thoroughly and make sure that your work is correct before continuing.

You will learn to identify the three types of circuits and what makes them series, parallel, or series-parallel circuits. You will also use formulas to determine the total voltage, current, and resistance of the circuits. By using these formulas and Ohm's Law, you will be able to answer many questions for yourself.

Performance Objectives

a. Identify series, parallel, and series-parallel circuits, from schematics. 100% accuracy required.

b. Measure d.c. voltage in a series circuit, using the multimeter. Reading must be within two graduations of the correct value.

c. Measure d.c. voltage in a parallel circuit, using the multimeter. Reading must be within two graduations of the correct value.

d. Measure d.c. voltage in a series-parallel circuit, using the multimeter. Reading must be within two graduations of the correct value.

e. Measure d.c. current in a series circuit, using the multimeter. Reading must be within two graduations of the correct value.

f. Measure d.c. current in a parallel circuit, using the multimeter. Reading must be within two graduations of the correct value.

g. Measure d.c. current in a series-parallel circuit, using the multimeter. Reading must be within two graduations of the correct value.

h. Measure single, multiple or total resistance values in a

series circuit, using the multimeter. Reading must be within the indicated tolerance.

i. Measure individual, branch or total resistance values in a parallel network, using the multimeter. Readings must be within the indicated tolerance.

j. Measure individual, branch or total resistance values in a series-parallel circuit, using the multimeter. Reading must be within the indicated tolerance.

Equipment Required

Multimeter (VOM), soldering equipment, resistor-diode trainer, and fuse-switch block.

[Following are reproductions of actual Workbook pages.]

OBJECTIVE a. Identify a series circuit in a schematic diagram.

READING ASSIGNMENT: Pages 46-48, frames 3-1, to 3-9; pages 70-76, frames 4-27 through 4-67 of Basic Electricity. Complete self-test 2, pages 76 and 77.

TRAINER CHECK 25. Which of the circuits below is not a series circuit?

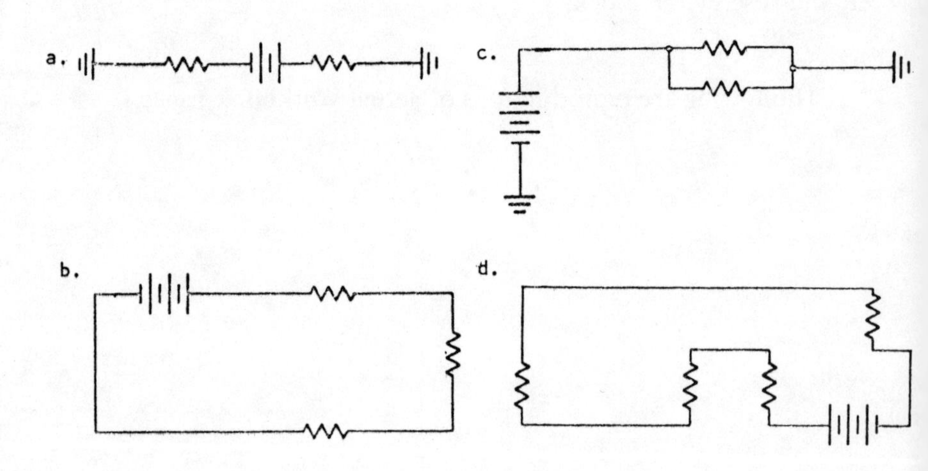

Figure 1.

Remember, a series circuit has only one path for current to flow. There may be one or more resistors connected end to end, but it remains a series circuit with only one path.

OBJECTIVES b, e, and h. Measure and calculate series voltage, current, and resistance.

STEP 1. Install jumper wires on the resistor-diode trainer to complete a series circuit. The board is numbered 1 through 8 across the bottom and A through G down the right-hand side of the board. Locate the number 3 on the bottom of the trainer board. You will use the resistors in row 3, A through G.

Connect a jumper wire across points 3B and 3C.

Connect a jumper wire across points 3E and 3F.

STEP 2. Set the VOM to measure resistance. If you are not sure how to do this, refer back to Project 1.

Measure the resistances at the indicated test points and record them below.

Test Points	Resistance
3A - 3B	________
3C - 3D	________
3D - 3E	________
3F - 3G	________

STEP 3. The resistance readings should be approximately

500K at resistor 3A - 3B
50K at resistor 3C - 3D
5 K at resistor 3D - 3E
50K at resistor 3F - 3G

Add all the resistances in STEP 3 above. Total____________

The formula for total resistance in a series circuit is

$$R_t = R_1 + R_2 + R_3 + R_4$$

R_t is total resistance, R_1 is resistance of resistor 1, R_2 is resistance of resistor 2, etc.

STEP 4. Measure the resistance from 3A to 3G. The meter reads________.

The formula $R_t = R_1 + R_2 + R_3 + R_4$ has proven itself. Your total was above 600K when you added the four resistors. When you measured the total between 3A and 3G, you also got approximately 600K.

STEP 5. Study the series circuit that you made on the trainer board. Notice there is only one path for current flow and that the sum of the individual resistors equals the total resistance.

Draw a schematic of a series circuit containing five resistors; label them R_1 through R_5. Connect these resistors to a 100_v battery. Show 5 amps flowing in the circuit.

STEP 6. Your circuit should be similar to the one in Figure 2.

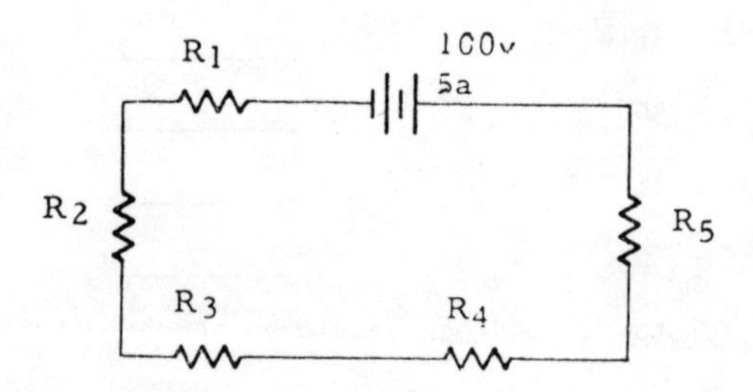

Figure 2.

INSTRUCTOR CHECK. Label R_4 with the amount of current flowing through it. Have your instructor check your work.

For a voltage source you will be using the low voltage power supply (LV) mounted in your work bench. This power supply is the unit located on the left hand side of the work bench. It consists of a voltage control, volt meter, amp meter, and two output terminals. Also, on the bottom of the unit there is an ON and OFF switch, plus a pilot light to indicate the ON condition of the power supply.

CAUTION: Leave low voltage power supply ON ONLY while making measurements.

STEP 7. Remove the jumper wires in Row 3. Connect the LVDC power supply to the Resistor-Diode board, positive to point 1A and negative to the fuse of the fuse-switch block. Connect a lead between the switch and point 1G of the board. Install jumper wires between point 4B and 4C and between 4E and 4F. Turn the LV power supply on and adjust to 10 VDC on the meter face.

STEP 8. Turn the LV power supply off and remove the jumper at the positive post of the LV power supply. Connect the VOM as an ammeter at this point and turn the LV power supply on. Turn on the circuit switch and record the current_______________.

You should get between .4 and .43 ma. Reconnect the jumper wire to the positive post and connect the ammeter at points 4B--4C, record the reading._________________. You should get .4 to .43 ma here too. Reconnect the jumper at points 4B-4C. Install the ammeter at points 4E-4F and record your reading__________. The reading is still .4 to .43 ma. Reinstall LVDC power supply, record your reading_________. This same reading of .4 to .43 ma proves that the current flow through a series circuit is the same throughout the circuit. Thus, the current flow that you measured at each point was the total current (I_t) for the entire circuit.

TRAINER CHECK 26. The current flow in a series circuit:

a. is the same throughout the circuit.
b. is divided by each resistor.
c. varies through each resistor.
d. is multiplied by the resistance.

STEP 9. Let us go one step further and prove one more fact about series circuits. In STEP 4 you proved the formula $R_t = R_1 + R_2 + R_3$ .... It follows then that the sum of the voltage drops across all the resistors must equal the applied voltage (voltage drop or difference in potential across the power source.) Thus the formula $E_t = E_{R1} + E_{R2} + E_{R3}$ ....

E_t = Total voltage

E_{R1} = voltage drop across resistor R_1.................

Make sure that all jumpers are connected as follows: positive post - 1A, 4B - 4C, 4E - 4F, 1G - fuse-switch block, and fuse-switch block - negative post.

Now apply 10 vdc to the resistor-diode board. Then, measure and record the voltage drop between 4A and 4G.

STEP 10. Measure and record the voltage drop between each of the test points below.

4A - 4B ______________________________

4C - 4D______________________________

4D - 4E______________________________

4F - 4G______________________________

Your voltage readings should be: 4AB = 4.5v, 4CD = .45v, 4DE = .45v , and 4FG = 4.5v.

STEP 11. Add the voltage drops from STEP 10. The total is __________.

The total should come close to 10v. This proves that the sum of the individual voltage drops across each resistor is equal to the applied voltage (the total voltage drop) of the complete circuit.

Turn off and disconnect the power supply and remove all jumper wires from the resistor-diode board.

STEP 12. A series circuit has ______Path (s) for current flow.

STEP 13. A series circuit has only one path for current flow. Write the formula for finding total resistance in a series circuit containing three resistors.________________________________.

To find total resistance of a series circuit use the formula

$$R_t = R_1 + R_2 + R_3.$$

TRAINER CHECK 27. From the circuits of Figure 3, choose the one that is a series curcuit.

a.

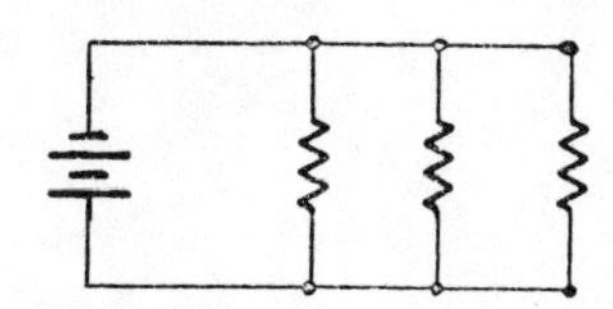

c.

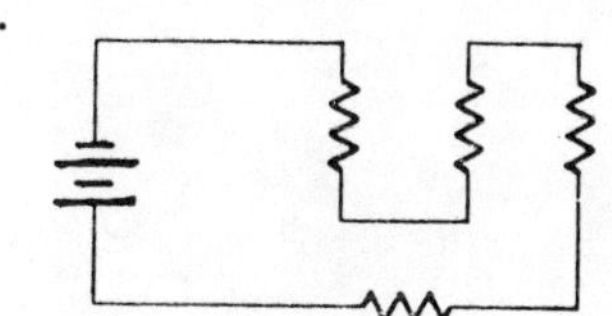

b.

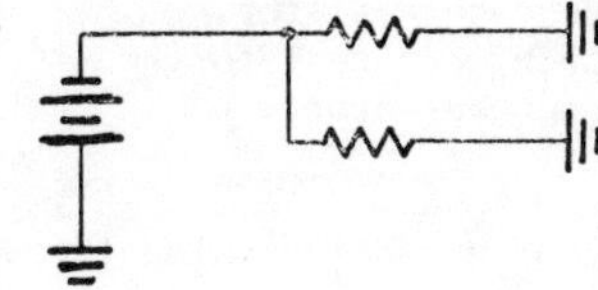

d.

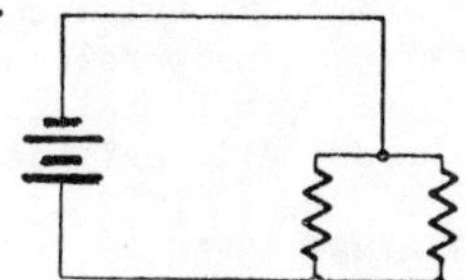

Figure 3.

TRAINER CHECK 28. Determine the total resistance of the circuit in Figure 4. Select the correct answer from those given.

a. 45K
b. 33K
c. 21K
d. 53K

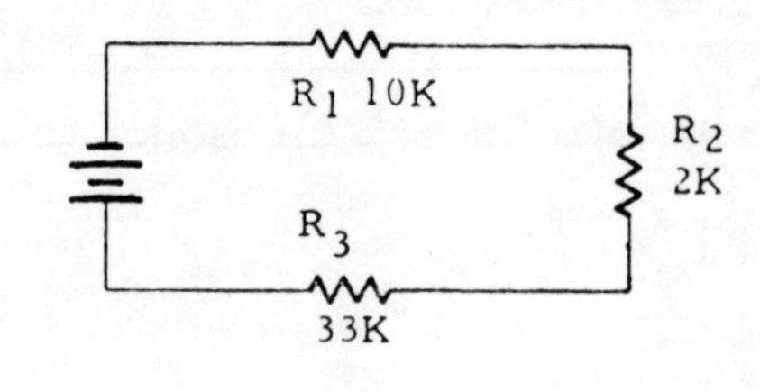

Figure 4.

TRAINER CHECK 29. Which of the statements below describes a characteristic of the series circuit?

a. The voltage drop is the same across each resistor.
b. The total current is the same through each resistor.
c. The total current varies through each resistor.
d. All resistors must be equal.

TRAINER CHECK 30. Determine the amount of current flowing through R_2 in Figure 5. Select the correct answer .

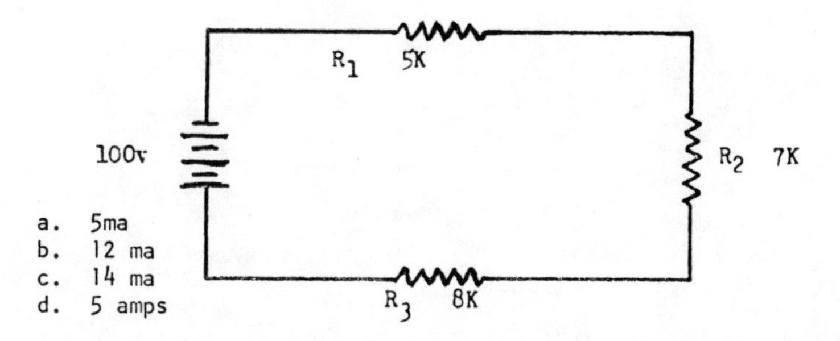

a. 5ma
b. 12 ma
c. 14 ma
d. 5 amps

Figure 5.

TRAINER CHECK 31. Determine the voltage drop across R_3 in Figure 5.

a. 4 v
b. 40 v
c. 12 v
d. 80 v

SUMMARY: In this portion of the project you worked only with series circuits. You connected resistors in series on the Resistor-Diode trainer; then you used the formula $R_t = R_1 + R_2 + R_3$..... With this formula, you proved that the total resistance is equal to the sum of the resistors in series. You also proved that the sum of the voltage drops across the resistors equalled the applied voltage (voltage drop at power source). By using an ammeter you proved that current flow in a series circuit remained the same throughout the circuit. Next, you will work with parallel circuits.

APPENDIX I

SAMPLE LEARNING UNIT

A learning activity guide that combines some newly created material with off-the-shelf technical material and adjunctive performance evaluation.

PERFORMANCE EVALUATION SET & LEARNER ACTIVITY GUIDE

POWER MECHANICS

Family: Auto Mechanics & Related Occupations.

Exit Level: Service Station Attendant & Related Occupations.

(915.867)

LEVEL I

CHASSIS LUBRICATION
PE 11-6

TASK 12 C. O. 1 & 2

NAME ____________________

DATE ____________________

LEARNER ACTIVITY GUIDE

PREREQUISITES: PE 3-1 and PE 11-1 through 11-5

OBJECTIVES: Given an auto to be lubricated, you will:

1. Use a service manual lube chart to locate and clean lubrication points in front suspension, steering linkages, drive and power lines, cables and linkages, etc.

2. Identify the proper tools and adapters and apply the specified type and amount of lubricant without dirt or foreign materials entering the system. Follow the lubrication chart directions for the specific make, model, and year of car.

3. Check lubricant level in differential, manual transmission, manual steering gear, and power steering reservoir. Identify proper lubricant.

4. Identify and lubricate to specifications, various under-the-hood lubrication points.

(Continued)

PROJECT ABLE

OVERVIEW: Most cars and trucks have lube points on the underbody which are exposed to rugged operating conditions. The steering and suspension systems, with ball joints and bearings, are the major underbody lube points. Careful servicing is important. While older vehicles are equipped with grease "fittings" for such joints, most new cars are now sold with pre-packed bearings. The servicing interval for most fitting-equipped points is from 1,000 to 4,000 miles. The recommended servicing interval for pre-packed bearings ranges from 12,000 miles to 36,000 miles (or from 12 months to 36 months). You must know that the method of lubrication is different for the two types. Greasing a pre-packed bearing like those equipped with standard fittings could ruin the bearing seals. Furthermore, a different type of grease is usually required. Chassis lubrication is one job you should not attempt without the careful supervision of the instructor or mechanic.

STUDENT-INSTRUCTOR CONTRACT OPTIONS:

- ☐ 1. Student-instructor conference.
- ☐ 2. Learning Unit #11-6.
- ☐ 3. Chek-Chart's Car Service, Chek-Chart Corporation, pp. 49-54.
- ☐ 4. Other--specify: ______________________.

EQUIPMENT: Tote-Tray #11-6 with lube chart manual, penetrating fluid, oil can with 10W30 oil, hand lubrication gun, adapters for pre-packed bearings, and assorted wrenches. Get some paper towels.

POWER MECHANICS
Family: Auto Mechanics & Related Occupations.
Exit Level: Service Station Attendant & Related Occupations.
(915.867)
LEVEL I

CHASSIS
LUBRICATION
PE 11-6

TASK 12 | C. O. 1 & 2

NAME ____________
DATE ____________

Pre Assessment

Instructions;

(1) Fill in name and date on the last two pages. When you have completed the performance evaluation, you will get one copy the instructor will file the other.

(2) Do the training check questions below and give answer card to instructor.

(3) Complete the performance evaluation under instructor's supe vision. He must see proof of your performance.

TRAINING CHECKS: T-T No. Z-11. The correct answer is **L**.
Start with number **17**.

17. Dirt must be removed from fittings and plugs

   a. to make a path for excess grease.
   b. to prevent foreign materials from entering bearing.
   c. to see the bearing.
   d. to present a neat appearance.

18. To remove the grease gun from a fitting after greasing the lube point

   a. unscrew fitting.
   b. pull straight off.
   c. break by moving up, down, or sideways.
   d. pull trigger and pop out.

19. Limited slip differentials can always be detected by

   a. checking drain plug for metal tag.
   b. checking manual for specifications.
   c. checking special type of grease in differential.
   d. rotating a rear wheel and observing opposite wheel.

20. The service interval for bearings with standard fittings and for pre-packed bearings is

    a. much longer for pre-packed bearings.
    b. determined by the mechanic.
    c. longer for the standard fitting equipped bearings.
    d. about the same.

21. The pressure gun

    a. can be used on pre-packed bearings by changing only the grease.
    b. can be used on pre-packed bearings with no modifications.
    c. should not be used on pre-packed bearings.
    d. should not be used unless the nipples are changed.

22. Limited slip differentials

    a. use a different grease than used in standard differentials.
    b. use the same grease furnished for standard differentials.
    c. are serviced the same as any other differential.
    d. require no special care.

23. The lubricant for manifold heat-control valves should be

    a. Door-Ease or silicon spray.
    b. penetrating fluid or similar lubricant.
    c. flake graphite.
    d. SAE 20 oil.

Identify the following (put a check mark next to the correct letter)

24. Standard nipple plug.

    a. ______
    b. ______
    c. ______

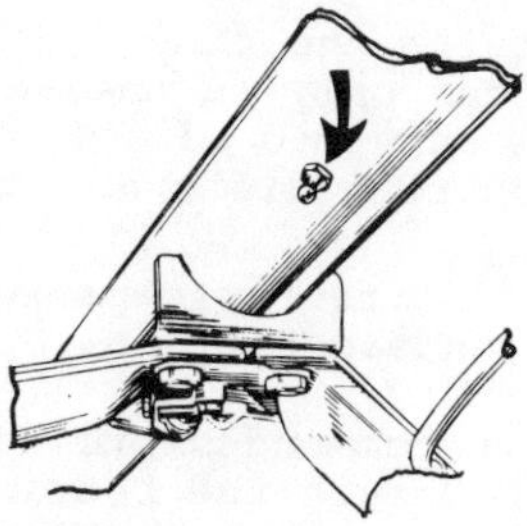

25. Pre-packed bearing plug.

    a. ______
    b. ______
    c. ______

26. Flush type plug.

    a. ______
    b. ______
    c. ______

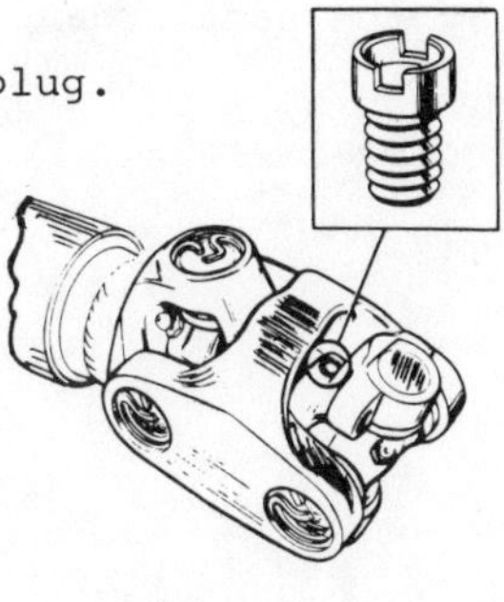

27. When the lubricant in a differential, steering reservoir, or transmission is <u>very</u> low, you should

   a. recommend the owner return it at a later time for service.
   b. recommend draining and refilling unit with new fluid.
   c. simply fill to proper level with specified lubricant.
   d. add gear grease.

28. Vehicles should be allowed to warm up indoors before greasing when the temperature approaches

   a. 0°F.
   b, -10°F.
   c. +10°F.
   d. +20°F.

29. When attaching grease gun to fitting,

   a. push straight onto fitting.
   b. touch lightly and apply grease.
   c. pull trigger and shove.
   d. place on angle and roll on.

30. Most new cars are sold with

   a. pre-packed bearings for most front-end lube points.
   b. standard grease fittings for most lube points.
   c. standard grease fittings for all lube points.
   d. standard nipple plugs for most lube points.

31. Standard fittings and pre-packed bearings

   a. require the same type of grease.
   b. differ only in the service interval.
   c. are serviced with the same tools and fittings.
   d. require a different type of grease.

32. Greasing either pre-packed bearings or bearings equipped with standard fittings.

   a. is recommended procedure.
   b. could ruin the bearing seals.
   c. requires essentially the same tools but different grea
   d. requires essentially the same grease but different too

STOP ____________________ INSTRUCTOR CHECK #1
initials

POWER MECHANICS
Family: Auto Mechanics & Related Occupations.
Exit Level: Service Station Attendant & Related Occupations.
(915.867)
LEVEL I

CHASSIS LUBRICATION
PE 11-6

TASK 12 C. O. 1 & 2

NAME ____________

DATE ____________

PERFORMANCE ACTIVITY
(Pre and/or Post Assessment)

UNIT OBJECTIVE 1:	Using a service manual lube chart, locate and clean lubrication points in front suspension, steering linkages, drive and power lines, cables and linkages, springs, etc.

A. ____________ year ____________ make ____________ model ____________ mileage

When was the vehicle last greased? ____________ miles ____________ date

B. What is the recommended lubrication service interval?

____________ miles ____________ months

Does the mileage or time interval indicate the need for greasing? ____________

C. How many plugs are listed? ____________

How many fittings are listed? ____________

What type of plugs or fittings are listed?

____

____

____

#33. Do all plugs and fittings require the same type of lubricant?

a. No b. Yes

D. Complete the following information:

Differential

Type ____________________

Service Interval ____________________

Lubricant ____________________

Is service required? ____________________

Transmission-Overdrive (or Automatic Transmission)

Type ____________________

Service Interval ____________________

Lubricant ____________________

Is service required? ____________________

Steering

Type ____________________

Service Interval ____________________

Lubricant ____________________

Is service required? ____________________

E. Raise vehicle following procedures listed in unit on lifts and jacks.

F. Prepare the plugs for greasing--do NOT grease until after the instructor check below.

What type plug or fitting is used? ____________________

Were the plugs changed? ____________________

Were the plugs originally of the pre-packed type? ________

UNIT OBJECTIVE 2:	Identify the proper tools and adapters and apply the specified type and amount of lubricant without dirt or foreign materials entering the system. Follow the lubrication chart directions for the specific make, model, and year of car.

A. Get the hand gun. It should be filled with the lubricant specified for pre-packed bearings. Is it the type of lubricant specified by the manual? _______________ Do NOT grease anything yet.

#34. Get the pressure gun. Does it have the type of lubricant specified for standard nipple-type fittings?

a. No b. Yes

NOTE: The pressure system has the wrong type of grease for pre-packed bearings. NEVER use the pressure gun on sealed pre-packed bearings. The pressure would break the seals--this could void the warranty.

NOTE: No student (10th, 11th, or 12th grade) is allowed to grease fittings without first having the job inspected by the instructor.

STOP ________________
initials

INSTRUCTOR CHECK #2:
Check written work. Check identification of fittings. Student must be able to identify pre-packed bearings. Make certain he has identified and cleaned all lube points. Check for limited slip differential. Have student identify plugs on differential and transmission. Have student demonstrate use of hand gun and pressure gun. Watch him perform. Make certain he keeps fittings and nozzle VERY clean. Have student demonstrate turning of wheels while greasing ball joints or king pins.

B. Do NOT attempt to grease a universal joint or drive shaft without instructor's assistance. Lubricate the first few points with instructor's help.

C. Lubricate all fittings and plugs as indicated on chart. Use proper lubricant.

UNIT OBJECTIVE 3: Check lubricant level in differential, manual steering gear, power steering reservoir, and manual transmission-overdrive unit.

Differential

A. What type of lubricant is specified for the standard differential? ______________ What type of lubricant is specified for the limited slip differential? ______________ (Check the service chart for some other make of car if both are not listed for the vehicle you are servicing.) Does the vehicle have a limited slip differential? ________________

B. Find and prepare plug--do NOT remove until checked by instructor.

STOP _______________ initials

INSTRUCTOR CHECK #3:
Have student remove plug, check level, and replace plug. Did student inspect for leaks and broken seals?

C. Is lubricant required? ___________ Fill only by permission of instructor.

NOTE: Do not lower car to ground until instructor checks plug.

Manual Transmission

#35. What type of lubricant is specified?

a. A.T.F.

b. SAE 90-140

c. SAE 10W30

d. SAE 30

A. Find and prepare plug.

NOTE: Do not remove fill plug until checked by instructor. Should the car you have been servicing have an automatic transmission, go to another vehicle for this part of the project.

STOP _______________ initials

INSTRUCTOR CHECK #4:
Have student remove plug, check level, and replace plug. Did student inspect unit for leaks?

B. Is lubricant required? ______________

Fill only by permission of instructor.

Steering Gear (units without power steering)

#36.. What type of lubricant is specified?

a. Chassis lube

b. A.T.F.

c. SAE 10W

d. SAE 90-140

A. Find and prepare plug.

STOP ________________ INSTRUCTOR CHECK #5:
initials — Have student loosen fill plug, check fluid level, and replace plug.

B. Is lubricant required? ______________

Do not add lubricant without instructor's or mechanic's permission.

Power Steering Reservoir

#37. What type of lubricant is specified?

a. A.T.F.

b. SAE 10W

c. SAE 10W30

d. SAE 90-140

A. Find and prepare cover or fill cap.

Some older cars with power steering have two separate lube points: (1) the power steering unit reservoir and (2) the steering gear box. In new vehicles, the power steering reservoir supplies the gear box with lubricant. Your instructor can explain this.

B. Remove cap and check level. Is lubricant required? ________

Fill only by permission of instructor.

UNIT OBJECTIVE 4: Identify and lubricate, to specifications, various under-the-hood lubrication points.

Manifold Heat-Control Valve

#38. What is the specified lubricant?

a. SAE 30

b. A.T.F.

c. Penetrating oil

d. SAE 90-140

A. Lubricate.

Throttle Linkage

A. What is the specified lubricant? ____________________

B. Point out lube points to instructor--from manual.

C. Lubricate.

Other Accessories

A. List four (4) other lubrication points listed in manual. (Points not covered in this project.)

1. ____
2. ____
3. ____
4. ____

STOP ________________ initials

INSTRUCTOR CHECK #6:
Check steps in power steering, manifold heat-control valve, throttle linkage, and "other accessories".

POWER MECHANICS

Family: Auto Mechanics & Related Occupations.

Exit Level: Service Station Attendant & Related Occupations.

(915.867)

LEVEL I

CHASSIS LUBRICATION
PE 11-6

TASK	C. O.
12	1 & 2

NAME ______________
DATE ______________
MAN–HOURS ______________
INSTRUCTOR ______________

CRITERION CHECKLIST

L	M	S
☐	☐	☐

Lubricates chassis.

U S

☐ ☐ 1. Uses a service manual lube chart to locate and clean lubrication points in front suspension, steering linkages, drive and power lines, cables and linkages, etc.

☐ ☐ a. Identifies the service requirements and interval for the various lube points.

☐ ☐ b. Locates all lubrication points.

☐ ☐ c. Cleans all foreign matter from fittings and/or plugs.

☐ ☐ d. Identifies pre-packed bearings.

☐ ☐ 2. Identifies the proper tools and adapters and applies the specified type and amount of lubricant without dirt or foreign materials entering the system. Follows the lubrication chart directions for the specific make, model, and year of car.

☐ ☐ a. Identifies proper lubricant.

☐ ☐ b. Uses hand gun for pre-packed bearings and universal joints.

☐ ☐ c. Connects and breaks connection properly with both hand and pressure guns.

☐ ☐ d. Lubricates without foreign materials entering system--keeps nozzle and fitting clean.

☐ ☐ e. Follows lubrication chart directions.

U S

☐ ☐ 3. Checks lubricant level in differential, manual transmission, manual steering gear, and power steering reservoir. Identifies proper lubricant.

☐ ☐ a. Checks for limited slip differential.

☐ ☐ b. Checks differential.

☐ ☐ c. Checks manual transmission.

☐ ☐ d. Checks manual steering gear.

☐ ☐ e. Checks power steering reservoir.

☐ ☐ 4. Identifies and lubricates, to specifications, various under-the-hood lubrication points.

☐ ☐ a. Manifold heat-control valve.

☐ ☐ b. Throttle linkage.

☐ ☐ c. Others.

☐ ☐ 5. Performs tasks in an appropriate amount of time.

APPENDIX J

SAMPLE LEARNING UNIT

A minimal learning activity guide that is entirely dependent on off-the-shelf technical materials and adjunctive (formative) performance evaluation to structure and organize the student's activities.

Learning Activity Guide

TITLE: Use of multimeter in d.c. measurement.

OBJECTIVES: Given any multimeter and the manufacturer's operating instructions manual, you will demonstrate the correct procedures for measuring d.c. voltage, current, and resistance over the full range of the meter, using simple resistive d.c. circuits.

APPLICATION: The ability to use multimeters properly and to make accurate d.c. current, voltage, and resistance measurements is an essential skill to a technician working in an electronics-based technology. Such instruments are commonly used in trouble shooting, for example, to check values against specifications for various test points in a circuit.

EQUIPMENT, TOOLS, MATERIALS: Multimeter, manufacturer's instruction manual, resistance trainer board, d.c. power supply, soldering equipment, and standard hand tools.

TECHNICAL TERMS:

Multimeter	Polarity	Calibration
Function	Current	Tolerances
Range	Voltage	Sensitivity
Accuracy	Resistance	

Pre- or Post-Performance Check

Step 1: Answer the following questions.
Note: Use the self-scoring response card, L-1#Z -11. L indicates the correct answer for each question.

1.1 The control that determines the type of measurement to be made (voltage, current, resistance) is called the
- a. range selector.
- b. polarity switch.
- c. function selector.
- d. shunt selector.

1.2 The term applied to meter scales which defines the minimum and maximum reading of that scale is called
- a. calibration.
- b. range.
- c. limits.
- d. sensitivity.

1.3 When making a measurement, if the meter reads down-scale, you can correct this condition by either reversing the probe leads or by changing the position of the
- a. function switch.
- b. a.c.-d.c. switch.
- c. polarity switch.
- d. range switch.

1.4 Most meters have a screw showing directly below and in line with the pointer. This screw is used to
- a. adjust the mid-screw reading.
- b. set the mechanical zero.
- c. adjust the pivot bearing.
- d. set the full-scale reading.

1.5 Suppose the meter reads up-scale beyond the scale markings. You can correct this condition by
- a. changing the position of the polarity switch.

b. switching to a lower range.
c. adjusting the pointer zero.
d. switching to a higher range.

1.6 When making a resistance measurement, the probe leads should be connected between
a. output & common.
b. ohms & common.
c. output & ohms.
d. input & common.

1.7 The sensitivity of the meter is given as
a. volts per ohm.
b. ohms, full-scale.
c. ohms per volt.
d. ohms per division.

1.8 When making a resistance measurement, you first short the test probes and then calibrate the meter by adjusting the
a. zero adjust.
b. ohms adjust.
c. sensitivity adjust.
d. polarity switch.

1.9 When calibrating the meter for a resistance measurement, you set the pointer to
a. ∞ (infinity).
b. center scale.
c. fully counter clockwise.
d. zero ohms.

1.10 When you move the meter from a vertical (standing) position to a horizontal (lying) position and vice versa, you must
a. re-calibrate the ohms zero.
b. check the meter for a full-scale reading.
c. reset the mechanical zero.
d. check the battery voltage.

STOP! _____ INSTRUCTOR CHECK: Determine need for additional instruction and advisability of proceeding with performance test.
initials

Step 2: Obtain at least two different types of multi-function meters, locate the proper meter module. Get circuit materials specified by your instructor.

Step 3: Connect circuit #1 using the resistor R_1.
Note: Do not energize the circuit until Step 5.

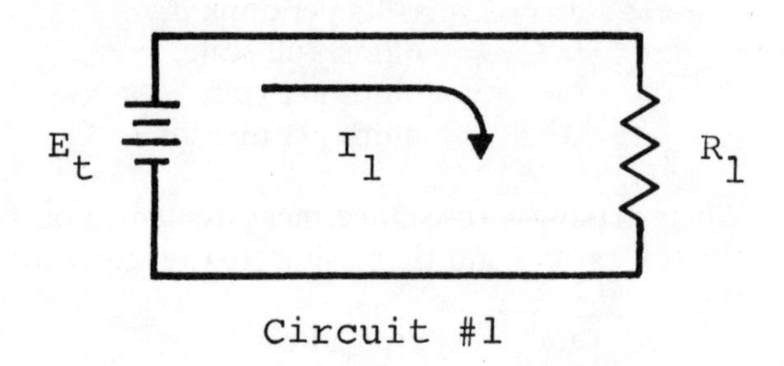

Circuit #1

Step 4: Connect a meter properly in the circuit to measure V_1.

STOP! _____ INSTRUCTOR CHECK: Do Criterion Checks #1 and 2 through 3-4 to insure that the student can operate the equipment properly. Check on variety of multi-function meters. Quiz student on ability to complete performance test.
initials

Step 5: Make the measurements required for Circuit 1 in the table below.

Step 6: Connect Circuit #2, adding the resistor R_2.

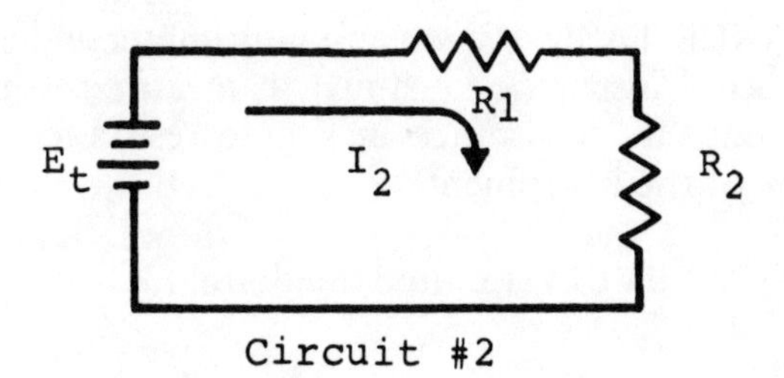

Circuit #2

Step 7: Use a different instrument and make the measurements required for Circuit #2.

Step 8: Connect Circuit #3, adding the resistor R_3.

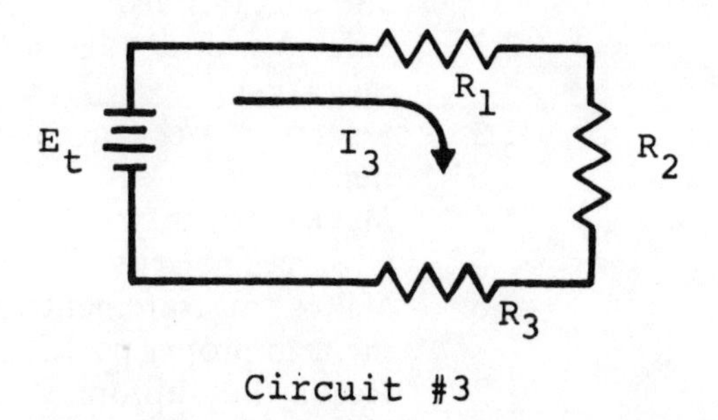

Circuit #3

Step 9: Make the measurements required for Circuit #3.

Circuit 1				Circuit 2				Circuit 3				
V_1	E_t	R_1	I_1	R_2	V_1	V_2	I_2	R_3	V_1	V_2	V_3	I_3

STOP! _____ INSTRUCTOR CHECK: Complete Criterion
initials Check list.

Instructor Criterion Check List

PERFORMANCE TASK: Given any multimeter and the manufacturer's operating instruction manual, can correctly measure d.c. voltage, current, and resistance in simple resistance circuits over the full range of the instrument.

PROFICIENCY LEVEL: Limited, Medium, Skilled.

SUB-TASKS:

- □ 1. Locates manufacturing equipment instructions or manual and follows directions for proper use of instrument.
- □ 2. Operates a variety of multi-function meters over the full range of such instruments.
  - □ 2.1 Makes required zero adjustments and calibrations.
  - □ 2.2 Selects proper function (100%–no mistakes).
  - □ 2.3 Selects proper range for initial reading.
  - □ 2.4 Makes proper circuit connection.
  - □ 2.5 Observes polarity.
  - □ 2.6 Makes measurements with instrument in proper position.
  - □ 2.7 Reads the appropriate scale within its accuracy.
  - □ 2.8 Makes measurements within three to five minutes.
  - □ 2.9 Observes standard instrument and circuit precautions.

APPENDIX K

SAMPLE GROUP-PACED LECTURE-DEMONSTRATION

A Basic Course in Radio Repair

There are times when it is more economical or more feasible to use a group-paced lecture-demonstration. Group instruction may be dictated by the lack of instructors or an overload of students, inadequate facilities or equipment, the impracticality of individualized learning materials, or perhaps the content and situation simply lend themselves best to lecture-demonstration techniques. If the lecture-demonstration makes full use of all the elements of an instructional system except self-pacing, group instruction can still be highly effective.

To be effective, a group-paced instructional system should have the following characteristics:

- Highly illustrated with all possible visual support such as slides, overhead, films, models, training aids, etc.
- Content is built around interim and terminal behavioral objectives that can be observed and measured.
- Content is organized as a step-by-step discussion and demonstration with interwoven student responses and performance followed by feedback knowledge-of-results to both the students and the instructor.

The details of a suggested model system are outlined below.

1. The instructor first demonstrates and discusses a task, concept, or rule, etc., with visual or model, covering at least a paragraph but no more than the equivalent of a page of text.

2. The instructor then immediately poses a question about the task or asks for performance of the task. The questions should call for application of knowledge or skill; not just the parroting of

the information just given, nor simply repeating a step just shown. The multiple-choice questions or statement of task to be performed can be shown by slide, overhead projector, or flip chart; or they may be in a prepared student handout that accompanies the demonstration. For instance, the task might be to calculate a mathematical value, to take a resistance reading with a multimeter, or just to answer a knowledge question.

3. Students respond by indicating choice with a response device of some sort. There are, of course, sophisticated devices and systems that could be used. However, simple expedients such as flash cards that indicate choice of an a, b, c, or d response, or even a show of hands for each response, can serve the purpose.

4. The instructor checks response devices and asks one or more students to explain their responses before telling them whether they are right or wrong.

5. The instructor follows up as needed to confirm the right and wrong responses.

6. Depending on class performance (as shown by their responses) and the criticality of the task, the instructor gives additional information on the same task, or moves on to next task. Flexibility of content should be built in. This way, the instructor can change pace to fit the situation, always trying to challenge the majority. The lecture outline should cover all terminal objectives and all the interim objectives leading to each terminal objective. The questions or tasks for students' responses at key points in the development should also be included in the outline. The instructor can skip from major objective to major objective as long as approximately 85 percent of the class can perform correctly without covering the intermediate sub-tasks. He should have his material prepared well enough to allow him to shift from fast track to slow track and back as needed.

7. The instructor (or assistant) keeps a record of missed responses (by response and by student), tallying the incorrect responses on a chart.

8. Drill and practice should be interspersed; not lumped at the end. Additional homework and tutorial help should be given to weak students. Reward good performance by decreasing amount of drill. Use good students as tutors for poor.

By adopting and adapting the techniques above, you can build an instructional system that does the following:

- Individualizes group instruction.
- Sets up a step-by-step, learning-by-doing situation, as opposed to passively listening and watching (dozing).
- Forces student participation by calling for active responding, and by asking for explanation of response before knowing whether they are right or wrong.
- Simulates, as closely as possible, the real job or skill being learned.
- Challenges the skill and flexibility of instructor to a high degree.
- Constantly challenges the student by making him perform continuously.
- Creates a self-correcting, adaptive instructional system. Because of its flexibility, it can more closely match the needs of both the group and the individual. The content and organization is constantly being evaluated against student performance and subsequently revised. Revision is based on performance data—both on-the-spot revision by the instructor as he adapts to needs of the students during the lecture-demonstration, and long-term revision based on the data collected over a period of time. The goal is to continue to revise the presentation until 85 percent of the students can respond correctly to each of the questions or correctly perform each of the tasks called for during the lecture.
- Knowledge-of-results tells the student how he stands at all times—he can't kid himself.
- Knowledge-of-results tells the instructor where he stands at all times—he can't kid himself either.
- A record of each student's performance can be kept throughout the instruction. This is the opening wedge for meaningful student evaluation.
- A record of the system's performance (including the instructor) can be kept through several classes. This is the opening wedge for meaningful evaluation of the instructor and the instructional system.

Probable results:

- Better student motivation because of challenging, meaningful activity.
- Faster learning, or higher level of learning and retention in the same length of time.
- Students will be better prepared for transition from classroom to application of knowledge and skill in the workshop.
- The instructor will find his job more challenging and rewarding.
- There should be a continuing rise in the quality of instruction because the system is self-evaluating, and thus it is a dynamic process, not a static product.

The following pages contain a sample portion of a lecture-demonstration that has been developed as an instructional system. The model format illustrates how a lecture-demonstration can become a true instructional system. The sample further demonstrates how any communication medium (book, lecture, slide-tape presentation, etc.) becomes a closed-loop informational system when the systems concept is applied.

Lecture-Demonstration #8
Lesson Plan

TITLE: Voltage, Current, and Resistance in Series, Parallel, and Series-Parallel Circuits.

OBJECTIVES:

1. Identify a series circuit in a schematic diagram.
2. Measure and calculate total series voltage, current, and resistance.
3. Identify a parallel circuit in a schematic diagram.
4. Measure and calculate total parallel voltage, current, and resistance.
5. Identify a series-parallel circuit in a schematic diagram.
6. Measure and calculate total series-parallel voltage, current, and resistance.

INSTRUCTOR EQUIPMENT: Multimeter, Resistance Trainer, soldering equipment, slide projector, slide set #8.

TRAINEE EQUIPMENT: Multimeter, Resistance Trainer, soldering equipment, individual work sheets.

OUTLINE OF TEXT: This project has three parts. The first portion covers series circuits, the second portion covers parallel circuits, and the third portion is on series-parallel circuits. The information in this project marks the real beginning of electronics education; thus you must make sure of your results—that every student in your class does, in fact, meet every objective.

Be sure each can identify the three types of circuits and what makes them series, parallel, or series-parallel circuits and also use the mathematical formulas to determine the total voltage, current, and resistance of the circuits.

Script Outline

OBJECTIVE I: Identify a series circuit in a schematic diagram: First, let's check to see if you did your reading assignment. Can you identify a series circuit in a schematic diagram? See if you can answer the question on the screen correctly. Show your choice on the response wheel. I'll wait for you to decide on your choice—a, b, c, or d.
(Show Slide)

Slide

QUESTION #1: Which of the circuits below is *not* a series circuit?

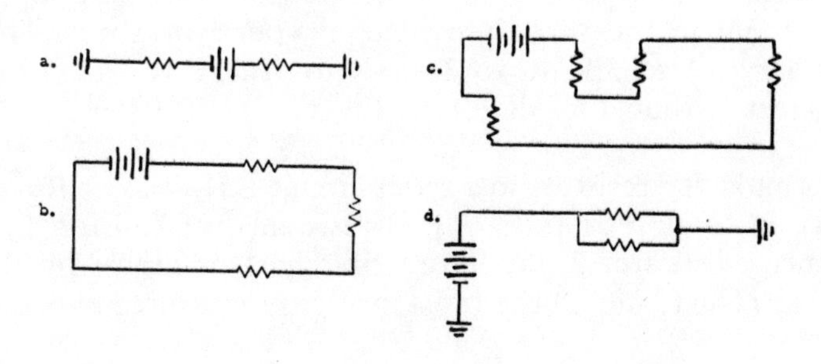

Mr. ___________, why did you pick "d" as the right answer?

Right, a series circuit has only one path for current to flow through, so "d" is the correct answer. There may be one or more resistors connected end to end, but it remains a series circuit with only one path, etc. (Further explanation and review may be needed if some students did not get the answer right. Ask them to explain their choice to find out why they were wrong and correct error.)

OBJECTIVE 2. Measure and calculate total series voltage, current, and resistance. Now, let's actually measure a series circuit voltage, current, and resistance on the trainer you have at your bench. First, install jumper wires on the resistor-diode trainer to complete a series circuit. The board is numbered 1 through 8 across the top and A through G down the right-hand side of the board. Locate the number 3 on the top of the trainer board. Use the resistors in row 3, A through G.

Connect a jumper wire to points 3B and 3C.

Connect a jumper wire to points 3E and 3F.

Demonstrate steps yourself. (Wait for each step to be completed.)

Now, set the multimeter to measure resistance. If you are not sure how to do this, just watch me as I do it.

Now, measure the resistances at the four indicated test points and record them in the space provided in your worksheet. The test points are: 3A to 3B; 3C to 3D; 3D to 3E; 3F to 3G. (Listed on work sheet or shown by slide.)

The formula for resistors in a series circuit is $R_t = R_1 + R_2 + R_3 + R_4$. R_t is total resistance, R_1 is resistance of resistor 1, R_2 is resistance of resistor 2, etc. (On work sheet or shown by slide or overhead.) Next, add all the resistances you measured above.

(Pause) You should get a total of 650 K. Now, measure from 3A to 3G. (Pause) The meter reads about 650 K also, doesn't it?

The formula $R_t = R_1 + R_2 + R_3 + R_4$ has proven itself. The total was about 650 K when you added the four resistors; and when you measured the total between 3A and 3G, you also got 650 K.

Now, study the series circuit that you made on the trainer board. Notice there is only one path for current flow and that the sum of the individual resistor equals the total resistance.

Using your worksheet, draw a schematic of a series circuit containing five resistors; label them R_1 through R_5. Connect these resistors to a 100-volt battery. Show 5 amps flowing in the circuit.

(Pause while they draw circuit on worksheet.)

The circuit you drew should look something like this.

(Show slide)

Slide

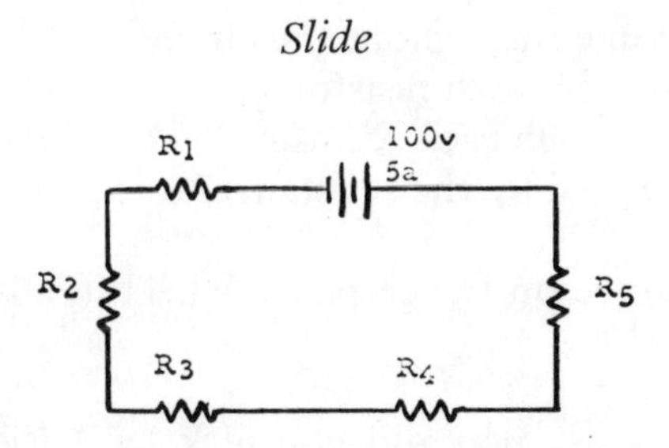

Now let's see what happens to the current in a series circuit. First, remove the jumper wires in row 3. Connect the power supply to the resistor-diode board at points A and 1A and points C and 1C.

Install jumper wires between points 4B and 4C; and between 4E and 4F. Now apply 20 vdc to the resistor-diode board.

Remove the jumper at point A of the power supply trainer. Connect the multimeter as an ammeter at this point and record the current on worksheet. (Pause for measurement. Demonstrate process.)

You should get between .8 and .85 ma. Reconnect the jumper wire at point A, connect the ammeter at points 4B–4C, and record the reading. (Pause) You should get .8 to .85 ma here too. Reconnect the jumper at points 4B–4C. Install the ammeter at points 4E–4F and record your reading. (Pause) The reading is still .8 to .85 ma. Reinstall the jumper at 4E–4F and connect the ammeter at point C of the power supply, and record your reading. (Pause) This same reading of .8 to .85 proves that the current flow through a series circuit is the same throughout the circuit. Thus, the current flow that you measured at each point was the total current (1_t) for the entire circuit.

Now, answer this. (Show slide)

Slide

QUESTION #2. The current flow in a series circuit

a. is the same throughout the circuit.
b. is divided by each resistor.
c. varies through each resistor.
d. is multiplied by the resistance.

Indicate your choice on the response wheel. (Pause until all have set response wheel.)

All right __________, why did you pick "a"? Right, the current flow is equal throughout a series circuit. (More explanation if needed to correct error.)

Let us go one step further and prove one more fact about series circuits. You proved the formula $R_t = R_1 + R_2 + R_3 \ldots$ It follows then that the sum of the voltage drops across all the resistors must equal the applied voltage (voltage drop or difference in potential across the power source). Thus, the formula $E_t = E_{R1} + E_{R2} + E_{R3} \ldots$ E_t = total voltage, E_{R1} = voltage drop across resistor $R_1 \ldots$

Make sure that all jumpers are connected at points A–1A, 4B–4C, 4E–4F, and 4G–C. Now, apply 20 vdc to the resistor board. Then, measure and record the voltage drop between 4A and 4G.

Now, measure and record the voltage drop between each of the test points. Record the voltages: 4A to 4B; 4C to 4D; 4D to 4E; 4F to 4G. (Pause while they measure. Demonstrate process.) The voltage readings should be: 4AB = 9v, 4CD = .9v, 4DE = .9v and 4FG = 9v.

Now, add the voltage drops from above and record total. The total should come close to 20v. This proves that the sum of the individual voltage drops across each resistor is equal to the applied voltage (the total voltage drop) of the complete circuit.

Remove all jumper wires from the resistor-diode board and disconnect the transformer.

Now, let's see how well you have understood what we have covered so far today. Answer these questions on series circuits. Use the response wheel. (Show slides and discuss each one as needed after calling on a student to give reason for choosing answer.)

Slide

QUESTION #3. From the circuits below, choose the one that is a series circuit.

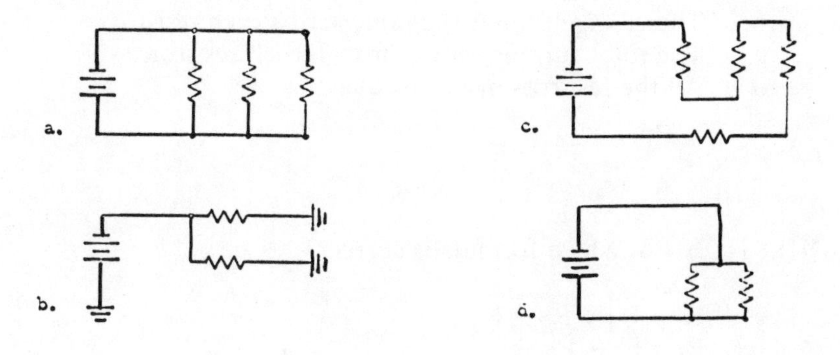

Slide

QUESTION #4. Determine the total resistance of the circuit below. Select the correct answer from those given.

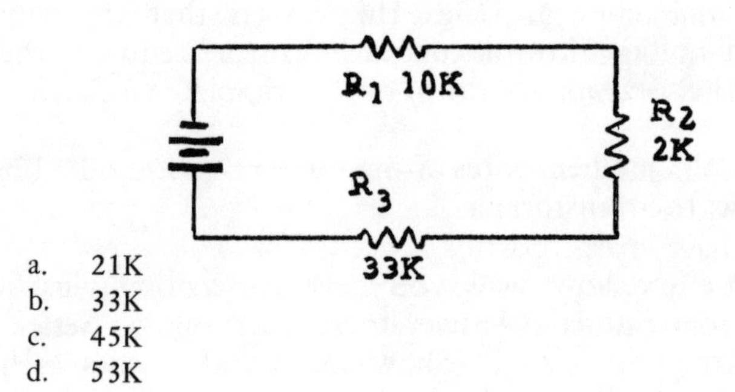

a. 21K
b. 33K
c. 45K
d. 53K

Slide

QUESTION #5. Which of the statements below describes a characteristic of a series circuit?

a. The total current is the same through each resistor.
b. The voltage drop is the same across each resistor.
c. The total current varies through each resistor.
d. All the resistors must be equal.

Slide

QUESTION #6. Which formula is correct?

a. $R_t = R_1 + R_2 + R_3 \ldots$
b. $E_t = E_1 + E_2 + E_3 \ldots$
c. Both a and b are correct.
d. Neither a nor b is correct.

SUMMARY: In this portion of the project you worked only with series circuits. You connected resistors in series on the resistor-diode trainer; then you used the formula $R_t = R_1 + R_2 + R_3 \ldots$ With this formula, you proved that the total resistance is equal to the sum of the resistors in series. You also proved that the sum of the voltage drops across the resistors equalled the applied voltage (voltage drop at power source), $E_t = E_{R1} + E_{R2} + E_{R3} \ldots$ By using an ammeter you proved that current flow in a series circuit remained the same throughout the circuit. Next, you will work with parallel circuits.

OBJECTIVE 3. Identify parallel circuits in schematic diagrams. All right, now let's get into parallel circuits. Parallel circuits, like series circuits, have characteristics of their own and may be identified by these characteristics. What do you remember about parallel circuits from your reading assignment?

Slide

QUESTION #7. Which circuit is *not* parallel?

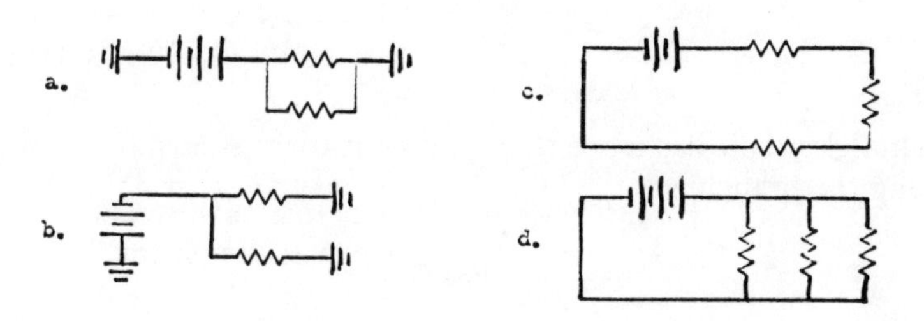

Why do you think "c" is the right answer, ____________?

Right, a parallel circuit is unlike a series circuit because it has two or more paths for current to flow through. (More explanation if there are errors.) One of the best ways to tell a series circuit from a parallel circuit is to trace the path of current flow. If there is one path, then you have a series circuit. If there is more than one path, it is a parallel circuit. For example:

Slide

Obviously, there is only one path for the circuit; so it is a series circuit. Note that the current flows from negative side of the battery to ground and from ground to the positive side. Here is another example of a series circuit.

Slide

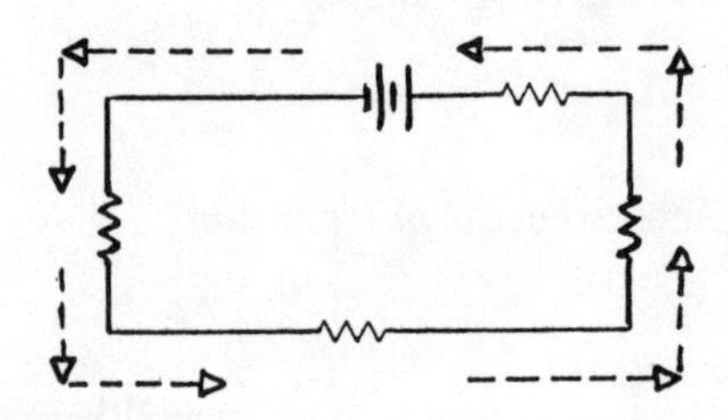

A parallel circuit has more than one path for the current. It divides among the branches.

Slide

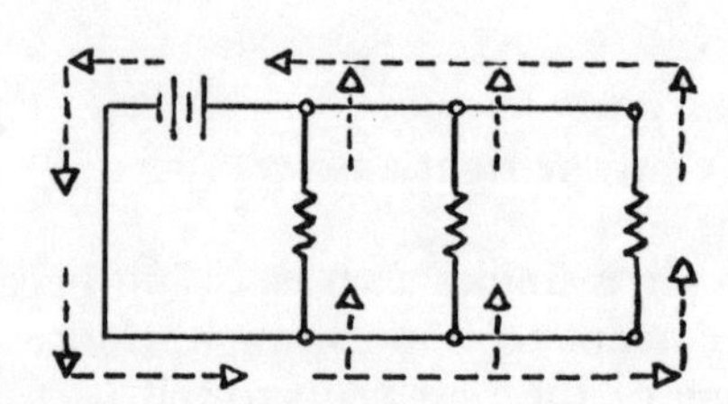

(This ends the sample, although the lesson is not complete here.)

APPENDIX L

SAMPLE OCCUPATIONAL READINESS RECORD

NOTE: The Occupational Readiness Record, which may be on a folding wallet-sized card, is straightforward and self-explanatory. As pointed out in the text, traditional letter grades are of no value in criterion-referenced testing and in learning to mastery. Note, however, that the criterion check list format does call for the instructor's expert judgment in rating the student's level of skill as limited (L), moderate (M) or skilled (S) in each critical task. Note also that the student is certified on only those tasks in which he has demonstrated the minimum level of competency stated in the performance objective.

OCCUPATIONAL READINESS RECORD

To the Employer:
This occupational readiness record is both an inventory of the training course content and a statement of the level of proficiency or achievement demonstrated by the graduate. Graduates can provide potential employers with more complete performance check lists which itemize in great detail the skills and knowledge in which they have demonstrated proficiency. It is recognized that persons working at the specified occupational level will function with direction and assistance from superiors. As a part of his training, the graduate has learned to expect appropriate instructions with each assigned task. Furthermore, the graduate understands that he lacks the authority and training to perform certain functions and operations. He will expect and seek supervision, assistance, and direction where appropriate. Note that the job tasks as identified are basic to the next higher or more sophisticated job

level. Work experience and further training may qualify the graduate for more complicated tasks, a new job title, and higher pay.

Key to Proficiency Code:
Level L: *Limited Skill*—does simple parts of task, using required tools, but requires instruction and supervision to do most parts of the job. Identifies part by name, knows simple facts about the job. Attainment of this rating is required to minimally qualify at the entry level.

Level M: *Moderate Skill*—requires help on some parts, but can use most tools and special equipment needed. Knows work procedures but may not meet minimum demands for speed or accuracy.

Level S: *Skilled*—understands operating principles and accomplishes all parts of task with only spot checks of finished work. Meets minimum demands for speed and accuracy.

All graduates receiving this document have satisfactorily demonstrated to the training staff their ability to work safely, understand and carry out instructions, and cooperate with other employees. This document also attests to their punctuality, reliability, and general work habits.

(Courtesy of Project ABLE, Quincy Public Schools and American Institutes for Research.)

JOB FAMILY: Auto Mechanics and Related Occupations
EXIT LEVEL: Service Station Attendant (915.867) and Related Occupations

Name____________________ Date____________________

Soc. Sec. No.____________________ Length of Training____________________

Certified by____________________ Title____________________

Comments____

L	M	S	
			Shop Safety
			Fire Safety
			Basic Mechanic's Handtools
			Automotive Hardware
			Automotive Terminology
			Identifies Customer Needs
			Cleans Service Area and Equipment
			Raises Cars With Floor Jacks and Combination Bumper-Frame Jacks
			Raises Cars With Twin-Post Hydraulic Lift
			Identifies and Replaces Defective Drive Belts
			Inspects Vehicle Lighting Circuit

L	M	S	
			Services Miniature Bulbs and Sockets
			Removes and Replaces Headlamps
			Identifies Common Spark Plug Deposits
			Cleans, Gaps and Tests Spark Plugs
			Removes and Replaces Spark Plugs
			Tests and Adjusts Tire Pressure
			Removes and Rotates Wheels
			Inspects Tires and Identifies Common Defects and Wear
			Mounts and Demounts Tubeless and Tube-Type Tires on Tire Machine
			Repairs Tubeless and Tube-Type Tires

L	M	S		L	M	S	
			Washes and Polishes Vehicles				Visually Inspects Cooling System Identifies Common Defects and Leak Points
			Tests Battery With Battery Hydrometer				Flushes and Fills Cooling Systems
			Inspects Batteries and Performs Minor Repairs				Tests Thermostats
			Cleans Batteries, Posts and Cables				Removes and Replaces Thermostats
			Removes and Replaces Batteries				Lubricates Body--Doors, Hinges, etc.
			Charges Batteries With Fast and Slow Charger				Identifies Specified Engine Oil, ATF and Lube Grease
			Inspects and Tests Radiator Pressure Caps				Checks Engine Oil and ATF and Fills to Proper Level
			Pressure Tests Cooling Systems				Determines Oil Lubrication and Filter Service Requirements
			Tests Antifreeze				Services Air and Gas Filters
			Identifies Common Hose Defects				Changes Oil and Oil Filter
			Removes and Replaces Hoses				Lubricates Chassis

The trainee has had limited experience in dispensing fuel, receiving credit and cash payments, and keeping records and inventory. On-the-job training required in these and other areas.

APPENDIX M

SAMPLE CAREER PLANNING OBJECTIVES AND STUDENT ACTIVITY GUIDE

If career planning and decision-making are to be effective, they must be systematically designed to achieve specific but limited objectives derived from the broader goals of the over-all guidance program. As pointed out in the final chapter, the task is feasible only through applying the same instructional systems development methods as used in preparing the training curriculum itself.

As always, developing crucial behavioral objectives is the key to the entire process. Thus, the sample performance objectives that follow are illustrative of how the global goals of a guidance program can be converted to explicit statements of desirable student activity.

The included sample individualized learning activity guide gives only 4 of 14 such career planning activities that are used in grades 9, 10, and 11. Similar but simpler activities are also carried out in grades 7 and 8. Only the four activities that come under the heading of "Matching Personal Credentials with Career Opportunities," which are the culmination of the planning exercises, are included. Note the other activities listed in the Table of Contents that precedes the learning unit.

Sample Career Planning Objectives

I. CHOOSING A CAREER
   A. Making a Realistic Choice

1. Self-evaluation
   a. Can evaluate own aptitudes and abilities with reference to broad occupational areas
   b. Can assess own interests as they apply to broad occupational areas
   c. Can evaluate own achievement level (academic standing) in relation to that of his peers for purposes of college entrance
   d. Can evaluate own educational achievement in relation to occupational areas
   e. Can evaluate own physical characteristics such as age, sex, or handicaps in relation to specifications set by employers for hiring
   f. Can assess own abilities and financial resources relative to the educational requirements for broad occupational areas
   g. Can assess own personality in relation to requirements of broad occupational areas
   h. Can evaluate potential occupational area choices in terms of his current life context (personal and parental aspirations, background, etc.)

2. Evaluation of world of work
   a. Can identify sources of information about occupations and occupational areas where vacancies exist; can extract pertinent information from the sources
   b. Can identify skill requirements for broad occupational areas
   c. Can describe or explain the concept of automation; can identify occupations for which skills and knowledge are likely to change in a relatively few years
   d. Can identify educational and achievement requirements for given occupational areas
   e. Can identify sources of post-high school training

f. Can select appropriate sources of financial assistance for post-high school training
g. Can identify professional or technical associations in which membership is required for performance in a given occupational area
h. Can identify the personality characteristics desirable for given occupations
i. Can assess the relative social status of various occupations
j. Knows where to find minimum job requirements for various levels within fields of work
k. Can evaluate the relative advancement possibilities associated with broad occupational areas

3. Combining knowledge of self and world of work—making tentative and general choices
   a. Can identify careers which have requirements and characteristics compatible with one's own aptitudes, skills, financial status, educational achievement, plans, etc.

B. Accepting the Consequences of Career Choice
1. Accepting the success and failure criteria
   a. Can identify the factors necessary for advancement in a given occupational area (skill-educational-interpersonal improvement)
   b. Can identify factors that can result in demotion or loss of job

2. Accepting career hierarchies and vertical mobility limits
   a. Can identify the occupational hierarchies associated with given occupational areas
   b. Can identify the requirements for moving to a higher position in a given occupation
   c. Can state own goals with respect to vertical mobility; can compare own goals with mobility limits of given occupational areas

3. Accepting the social status associated with an occupation
   a. Can identify the social demands which may be associated with given occupations (e.g., entertainment, dress)
   b. Can identify social roles required for successful performance of given occupations
   c. Can estimate the amount of time required for the social aspects of a given occupation
   d. Can evaluate own goals with respect to social status; can compare own goals with status limits of given occupational areas

4. Accepting duties and task requirements associated with an occupation
   a. Can identify duties and tasks associated with an occupation
   b. Can identify own abilities and personality characteristics with regard to duties and tasks

5. Demonstrating sensitivity to common satisfactions and dissatisfactions of an occupation
   a. Can assess the potential satisfactions and dissatisfactions associated with given occupations (e.g., a sense of accomplishment or no sense of accomplishment, recognition from subordinates and peers or unhappy co-worker relationships, high or low income, taking responsibility, being able to help others, routine or repetitive tasks, social status in community, dislike for superiors, undesirable working conditions and working hours, and opportunity to develop unique solutions to problems)
   b. Can evaluate own ability to perform a given job satisfactorily in spite of one or more major sources of dissatisfaction present

6. Accepting personal and family demands associated with an occupation
    a. Can identify and assess the personal sacrifices necessary for successful performance in a given occupation
    b. Can identify and assess hardships associated with given occupations that affect one's family

7. Accepting the roles of management, labor organizations, and government with respect to given occupations and to the economy
    a. Can assess the roles of labor organizations with respect to given occupations and the economy
    b. Can assess the roles of management or owners with respect to given occupations and the economy
    c. Can assess the roles of government with respect to given occupations and the economy

8. Accepting the environments, contexts, and settings of an occupation
    a. Can identify the environmental setting of a given occupation

C. Planning for Contingencies
1. Providing for technological change
    a. Can assess the extent to which technological change may downgrade or upgrade a given occupation by affecting the employment opportunities and occupational requirements

2. Accounting for social and economic trends
    a. Can identify social and economic trends and their effects on social and economic aspects of various occupations
    b. Can assess the future social and economic

status of a given occupation

c. Can identify alternative occupations whose training and experience are sufficiently similar to those of a given occupation that they may serve as alternate employment possibilities in case technological change eliminates a given occupation.

3. Providing for educational and occupational failure

a. Can identify alternative occupations in which the same training and experience is required, but in which the ability requirements are less strict

b. Can identify alternative careers which require the same abilities and interests, but which require less educational achievement

II. FORGING A CAREER

A. Preparing for a Career

1. Can identify realistic educational and training plans for a given occupational area

a. Can identify sources which provide information concerning the content of various educational and training courses

b. Can match occupational requirements with appropriate educational and/or training programs

c. Can identify alternative training routes to various occupational areas

d. Can identify length of time which must be committed to training for a given occupational area

e. Can estimate approximate cost of training in a given occupational area

2. Can assess ability to obtain financial aid for educational purposes

a. Can identify sources of financial aid

b. Can identify requirements or restrictions involved with obtaining scholarships and other financial aid

3. Learning critical requirements for a given occupation
   a. Can recognize the elements of a training course which are of critical importance to one's own specialty or area of interest
   b. Can identify non-required courses or extracurricular activities which can enhance one's skill and knowledge in critical areas

SAMPLE STUDENT CAREER PLANNING ACTIVITY GUIDE*

Table of Contents

Grade 9

*Prepared by American Institutes for Research, 135 N. Bellefield Avenue, Pittsburgh, Pennsylvania 15213 and Quincy Public Schools, Coddington Street, Quincy, Massachusetts 02169. Principal Author: Vivian (Hudak) Guilfoy as part of Project ABLE under Contract No. OE-5-85-019 with the Bureau of Research, Office of Education, U.S. Department of Health, Education, and Welfare.

INTRODUCTION

This is a very important year in your education. In the spring, you will be selecting a three-year high school program. Each activity in your Vocational Plan is designed to help make the process of decision making easier for you.

This year is important in a very special way: you must begin to narrow in on your decisions a little more than you have done up to this time. What you decide will effect your life at school next year at this time. Of course, it will be possible to make changes as you progress through school, but why not make the best possible decisions that you can--NOW.

Some of these activities may look familiar to you. This gives you a chance to compare how you are now with how you have been in the past. This year you will look at the data you have collected about yourself in the past, collect new data, and then USE THIS INFORMATION to help you plan ahead and make decisions. WHAT DECISIONS???

Take a closer look at the high school program planning form. It is important for you to think seriously about the type of work you want to do someday, and find out what the job requires and decide WHAT YOU CAN DO NOW to help you qualify for the jobs you want. The decisions you make about high school programs should help you reach your goals. These activities, if completed and considered carefully, will make the job easier for you and decision better for you.

Be sure to share this with your parents each step of the way. This is your chance to make decisions. You are the one who will profit by these activities.

GRADE NINE: MATCHING

Matching Personal Credentials with Career Opportunities

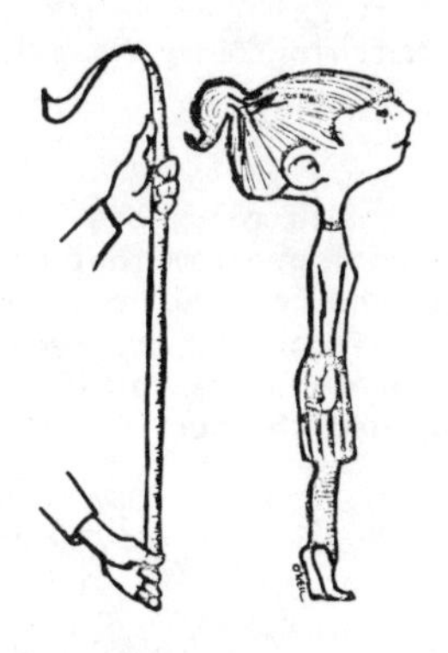

OBJECTIVES:

ANALYSE PERSONAL CREDENTIALS AND PREFERENCES BY COMPLETING A PERSONAL ANALYSIS.

COMPARE INFORMATION ABOUT SELF WITH INFORMATION ABOUT JOBS STUDIED.

CODE PERSONAL ANALYSIS DATA.

COMPARE CODED INFORMATION WITH MANY OCCUPATIONS.

COMPLETE AN OVERALL MATCHING SUMMARY.

OVERVIEW: During the earlier part of the year, you were asked to keep a detailed record of your achievements, your marks, your goals, and your interests. Now, even though some of these things may have changed, we ask that you go back to the pages of the Kit which cover Self Evaluation and think very carefully about what YOU as a person really look like. In order to make decisions about your future, you will need up-to-date, complete information about your abilities, interests, hopes and achievements. You will also need information about many jobs. You will compare this information about yourself with the information you discovered about jobs. This process will help you make a career decision which is appropriate for you.

LEARNING EXPERIENCES:

1. Complete the Personal Analysis Summary, page 52. Be sure to give the most up-to-date information about yourself.

2. Match your credentials with educational and vocational opportunities using the form on page 56. To do this, use your Occupational Analysis, p. 44 to 49 and your Personal Analysis, p. 52 to 55.

3. Take the information from the Personal Analysis, p. 52 to 55 and put it into an easy form called the Coded Personal Checklist, p. 59.

4. Compare your Coded Personal Checklist with any job you select in the Project ABLE Occupational Analyses Book. Your counselor will make this book available to you. Use the form on page 60.

5. Complete an Overall Matching Summary, p.61.

SUMMARY:

You have compared yourself with many jobs. Now you should be ready to select a high school program that matches your career plans. We hope that your choice will be a good one. Do not forget that through this guidance program you have also learned HOW TO MAKE DECISIONS. When you are faced with a new situation or want to change your mind, use the decision making process you learned in this program.

Grade 9: A Personal Analysis

The following questions are designed to help you identify your personal characteristics. You should use all of your Self Evaluation data as you answer each question. When you complete this Personal Analysis, you will have a summary description of your "credentials" and preferences. Then, in a next step, you will be able to compare yourself to the opportunities and requirements of the various jobs you have studied.

GENERAL

1. What kinds of tasks do you think you might like in a job?

____
____
____
____
____
____
____

2. In what industries would you prefer to work (see your Goal Checklist)

____
____
____

3. Do you expect to have one job or would you prefer to have several different jobs during your working lifetime?

_____only one job _____several jobs

4. Would you mind working at odd hours (nights, or in special seasons for example) or would you only accept a regular eight hour working day?

_____regular 8 hour day _____ any working hours

5. Estimate what you expect your average salary to be:

____$3000 to 4999	____$11000 to 14999
____ 5000 to 6999	____ 15000 to 19999
____ 7000 to 8999	____ 20000 to 49999
____ 9000 to 10999	____ 50000 and above

EDUCATIONAL AND TRAINING REQUIREMENTS

6. What high school courses are you planning to take? (Check Goal Checklist)

____Business Education	____General Woodworking
____College Preparatory	____Graphic and Commericial Art
____Computer Data Processing	____Health Occupations
____Electro Electronics	____Home Economics
____Foods Preparation	____Metals and Machines
____General Piping	____Power Mechanics

7. What types of education and training are you planning? (Check Goal Checklist)

_____High School
_____On-the-Job Training
_____Apprentice Training
_____Post HS Tech/Trade
_____Business College
_____Junior College
_____College
_____Graduate School

SPECIAL REQUIREMENTS

8. Complete the following items relating to your health:

How tall are you? ___ft. __in.
Do you wear glasses? ___Yes __No
How is your hearing? ___Good __Fair __Poor
How is your general health? ___Good __Fair __Poor

9. Do you think you might ever like to join a union? ___Yes ___No

10. Do you think you might like to join a professional association?

___Yes ___No

WORKING CONDITIONS

11. Which of the following conditions do you feel you could tolerate?

a.___working indoors
b.___working outdoors
c.___working in hot temperatures
d.___working in cold temperatures
e.___working in wet, humid conditions
f.___working where much noise or vibration is present
g.___work involving hazards and the risk of bodily injury
h.___working where fumes, odors, dust are present
i.___working where ventillation is poor
j.___other (specify):______________________________
k.___other (specify):______________________________

APTITUDES

For questions 12 through 19, check the three highest aptitude percentiles from your Differential Aptitude Test results.

12. ___Verbal
13. ___Numerical
14. ___Abstract
15. ___Spatial
16. ___Clerical
17. ___Mechanical
18. ___Language Usage: Spelling
19. ___Language Usage: Grammar

PHYSICAL DEMANDS

20. Which of the following types of physical activity do you feel you could tolerate (stand) as a regular part of the job?

a.___very heavy work
b.___climbing and/or balancing
c.___stooping, kneeling or crawling
d.___reaching, handling
e.___careful talking, hearing
f.___seeing, closely observing

INTERESTS

21. Do you enjoy doing things which involve things and objects? __Yes __No
22. Do you enjoy meeting and working with others? __Yes __No
23. Do you enjoy doing the same tasks daily? __Yes __No
24. Do you enjoy doing something which directly helps or gives service to others? __Yes __No
25. Do you enjoy helping to build the confidence or self-respect of others? __Yes __No
26. Do you enjoy trying to express your ideas or getting them across to other people? __Yes __No
27. Do you enjoy work of a scientific or technical nature? __Yes __No
28. Do you enjoy thinking up new ideas or ways to do things? __Yes __No
29. Do you enjoy working with machines, equipment, mechanical things? __Yes __No
30. Do you enjoy making products, things? __Yes __No

TEMPERAMENTS

31. Do you prefer doing a lot of different things which change frequently? __Yes __No
32. Do you like repeating the same job tasks? __Yes __No
33. Do you prefer to work under the direction of someone else? __Yes __No
34. Do you like to plan or direct the activities of others? __Yes __No
35. Do you like to work in close cooperation with others? __Yes __No
36. Do you prefer working alone? __Yes __No
37. Do you like to convince others or influence their ideas? __Yes __No
38. Do you like to make quick judgments under pressure? __Yes __No
39. Do you like to make decisions based on observation? __Yes __No

40. Do you like to work at tasks involving many facts and figures? __Yes __No

41. Do you like to help people put their personal ideas and feelings into words? __Yes __No

42. Do you like to do a great deal of precise and accurate work? __Yes __No

RELATION TO DATA, PEOPLE, THINGS

43. Do you prefer work which involves:

____people ____ideas (data) ____things

Matching Personal Credentials with Educational And Vocational Opportunities and Requirements

You have completed an Occupational Analysis and a Personal Analysis. Now you have the opportunity of comparing your own goals, achievements, and interests with the requirements of each occupation you have studied.

1. The left column tells you the kinds of questions you have asked about yourself and asked about the job you studied. The next column, titled OA and PA Item Number tells you which question in your Occupational Analysis (OA) and in your Personal Analysis (PA) answers the question. In the center column, titled MATCH, check whether the answer in your Personal Analysis which is the SAME as or DIFFERENT from the answer in the Occupational Analysis which describes the job.

For Example: To compare the educational requirements of a job with your own educational plans, compare Item #6 on the OA with Item #6 on your PA. If your plans MATCH with the job requirements, then check "same" next to the item number. If they do not, check "different."

2. Some topics have more than one question related to it. When you have matched yourself with jobs for all the questions connected with a topic, (for example, education and training requirements), complete the SUMMARY - this is where you can add special comments. Complete the entire process for each job you have studied.

This will provide important clues about the vocational goals you will set for yourself in the future and will help you select a high school program of studies which is best for YOU.

TOPIC	OA AND PA ITEM #	MATCH		SUMMARY
		Same	Different	
General	1			
	2			
	3			
	4			
	5			
Education and Training Requirements	6			
	7			
Special Requirements	8			
	9			
	10			
Working Conditions	11-a.			
	11-b.			
	11-c.			
	11-d.			
	11-e.			
	11-f.			
	11-g.			
	11-h.			
	11-i.			
	11-j.			
	11-k.			

TOPIC	OA AND PA ITEM #	MATCH		SUMMARY
		Same	Different	
Aptitude	12			
	13			
	14			
	15			
	16			
	17			
	18			
	19			
Physical Demands	20-a.			
	20-b.			
	20-c.			
	20-d.			
	20-e.			
	20-f.			
Interests	21			
	22			
	23			
	24			
	25			
	26			
	27			
	28			
	29			
	30			
Temperaments	31			
	32			
	33			
	34			
	35			
	36			
	37			
	38			
	39			
	40			
	41			
	42			
Relation to Data, People, Things	43			

WRITE IN ANY ADDITIONAL COMMENTS OR OBSERVATIONS YOU HAVE AT THIS TIME ABOUT THE MATCHING YOU HAVE JUST COMPLETED:

Coding My Personal Analysis Data

DIRECTIONS:

In order to put the information from the Personal Analysis (pages 52 to 55) into an easy form, which allows you to make a quick comparison with job information, do the following:

1. Look at the answers to items 1 - 5 on your Personal Analysis, p. 52. Write them in the left hand column of the Coded Personal Analysis Checklist, p. 59.

2. Where you answered YES, or said you could tolerate something, for questions 6 through 43 on the Personal Analysis, put an x next to that item on the Coded form under:

   HIGH SCHOOL COURSE OF STUDY
   TOTAL EDUCATION REQUIREMENTS
   CONDITIONS
   PHYSICAL DEMANDS
   INTERESTS
   TEMPERAMENT
   RELATION TO: Data, People, Things

3. Under APTITUDES, put an x next to your three highest percentiles, using the Differential Aptitude Test results.

4. Now, take the Occupational Analysis books and compare yourself with all the jobs in which you may be interested. Ask your counselor for this book and complete the summary on page 60.

CODED PERSONAL ANALYSIS CHECKLIST - GRADE NINE

1. JOB DUTIES:

2. INDUSTRIES:

3. OUTLOOK:

4. HOURS:

5. WAGES:

HIGH SCHOOL COURSE OF STUDY:	
1 Business Education	
2 College Preparatory	
3 Computer Data Processing	
4 Electro Electronics	
5 Food Preparation	
6 General Piping	
7 General Woodworking	
8 Graphic & Commercial Art	
9 Health Occupations	
0 Home Economics	
X Metals & Machines	
Y Power Mechanics	

TOTAL EDUC/TRNG REQUIREMENTS:	
1 HS Grade Completed: 10	
2 11	
3 12	
4 13 & 14	
5 On-The-Job Training	
6 Apprentice Training	
7 Post-HS Tech/Trade	
8 Business College	
9 Jr College (2 yr)	
0 College Graduate	
X Graduate School	

SPECIAL REQUIREMENTS:

CONDITIONS:	
1 Inside	
Outside	
Both	
2 Cold temp	
3 Hot temp	
4 Wet, humid	
5 Noise, vibration	
6 Hazards	
7 Fumes	
Odors	
Toxic	
Dust	
Poor ventilation	
8 Other	

APTITUDES:	
1 Verbal	
2 Numerical	
3 Abstract	
4 Spatial	
5 Clerical	
6 Mechanical	
7 Spelling	
8 Grammar	

PHYSICAL DEMANDS:	
1 Strength	
2 Climbing, balancing	
3 Stooping, kneel-ing, crawling	
4 Manual dexterity	
5 Talking, hearing	
6 Visual acuity	
7 Other	

INTERESTS:	
1 Dealing with things & objects	
2 Business contact with people	
3 Routine, system	
4 Social welfare	
5 Prestige, esteem	
6 Communication of ideas	
7 Science & technology	
8 Abstraction, creativity	
9 Machines, procedures	
0 Tangible results	

TEMPERAMENT:	
1 Varied duties, frequent change	
2 Repeated, set procedures	
3 Matching specified instructions	
4 Directing & planning for others	
5 Working with others	
6 Working alone & apart	
7 Influencing other people's ideas	
8 Risks, unexpected events, emergencies	
9 Making empirical judgments, decisions	
0 Analyzing facts and figures	
X Interpreting personal feelings	
Y Precision, accuracy	

RELATION TO:	
1 Data:	
2 People:	
3 Things	

HOW DO I COMPARE WITH JOB REQUIREMENTS - GRADE NINE

You now have an opportuntiy to compare your Coded Personal Analysis Checklist with a completed Occupational Analysis (ABLE/GUIDANCE/FORM G-1), for as many jobs as you wish. Summarize the results for each comparison you make:

1. Write in the name of the Vocational Area and the Job Title in the space provided.

2. Next to each topic listed, write in the number of ways your "credentials" are the same and the number of ways they are different from the job "credentials", and then summarize.

VOCATIONAL AREA: JOB TITLE:				VOCATIONAL AREA: JOB TITLE:		
TOPIC	# WAYS SAME	# WAYS DIFF.	SUMMARY	# WAYS SAME	# WAYS DIFF.	SUMMARY
General						
Education and Training Requirements						
Special Requirements						
Working Conditions						
Aptitudes						
Physical Demands						
Interests						
Temperaments						
Relation to Data, People Things						

OVERALL MATCHING SUMMARY - GRADE NINE

Using the results of the Matching process for all occupations studied, answer the following questions:

1. List the jobs in which you are still interested and give the reason(s).

Job	Reason for Interest

2. List the jobs that you feel are no longer suitable to consider and give the reason(s).

Job	Reason for Disinterest

3. What more would you like to learn about yourself or about jobs?

About You	About Jobs
______________________	______________________
______________________	______________________
______________________	______________________
______________________	______________________
______________________	______________________
______________________	______________________

4. Do you plan to continue the same course of study in high school that you stated in your Goal Checklist earlier this year? _____Yes _____No

   If you answered Yes, What is it?______________________________

   If you answered No, what changes do you think you would like to make?

   ____

   ____

   ____

   ____

   ____

   ____

   ____

APPENDIX N

CLASS MANAGEMENT SYSTEM

As pointed out in the chapter on "Implementing and Field Testing the System," individualized learning requires the instructor to assume a new role by becoming (with considerable help from students) a manager of the learning process and environment as well as personal tutor and counselor. Actually, the students' role change may be even more extreme as they assume more responsibility for their learning. The shift in responsibilities calls for a sharp change in attitudes as well as duties on the part of both the students and the instructors. Obviously, classroom procedures must also change considerably. And, unless the changes are carried out systematically, the resulting chaos will frustrate the best intentions of instructors and students.

The obvious need for the instructor to have a high degree of knowledge and acceptance of his new role is closely paralleled by the students' need for familiarization and practice in their new role. In particular, the prospect of assuming a major portion of the responsibility for his own progress, and that of his classmates, may be unsettling or even alarming to some students. Therefore, each student must have ample opportunity to become thoroughly familiar with these new classroom procedures before he is put on his own. The introductory process should be gradual at the outset. In fact, all the students should work through the first few learning units as a group to become acquainted with the steps involved. The sequence of events that typically occurs in an instructional system is diagrammed in the chart on page 326. Providing each student with a similar step-by-step, activity flow chart is very helpful.

INDIVIDUALIZED LEARNER ACTIVITY PROCESS WITHIN AN INSTRUCTIONAL SYSTEM

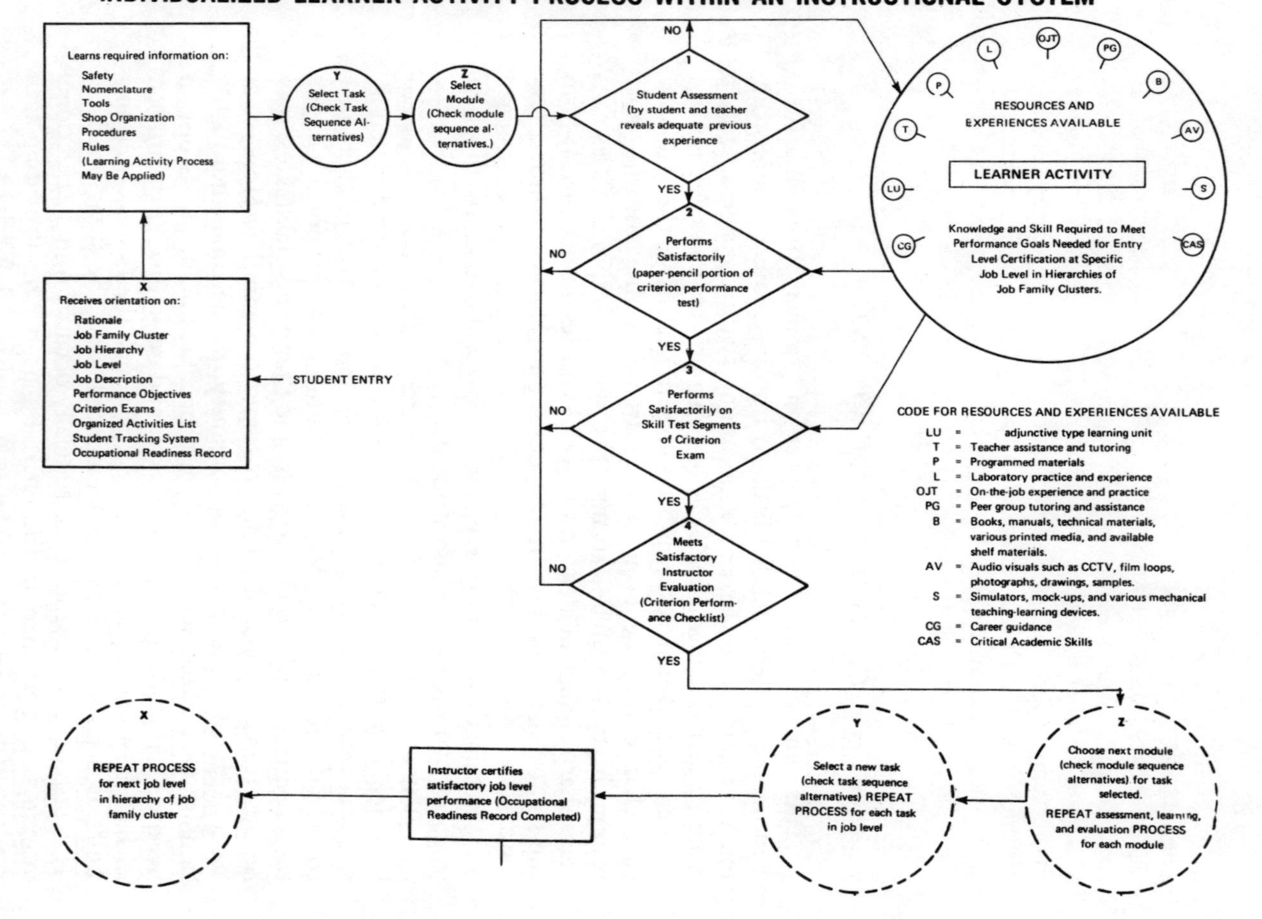

Another class management device that has proven to be important to the successful conduct of individualized learning is a "class status board." The basic function of a status board, such as that shown in the accompanying figure, is to provide the students and the instructor with a continuous graphic display of performance and progress. A glance at the status board instantly tells the instructor how well each student is progressing. More importantly, each student knows immediately what he has done and what he must do next. In addition, the instructor can also see immediately what he must do next. In effect, a properly designed board graphically displays the status of the learning contract between the students and the instructor.

Different colored tags show at a glance the status of every student:

Red tags show present physical location (absent, office, library, nurse, etc.) or which learning unit the student is presently working on.

Green tags indicate completion of the learning unit, successful completion of the performance test, and competency certification by the instructor.

Yellow tags show that the student has completed the learning unit but has not yet been given the performance test by the instructor.

Blue tags indicate the student by-passed the learning unit but demonstrated competency by successfully completing the performance test.

Because the students are largely responsible for their own learning progress, a major part of their own evaluation, and much of their program management, each student should also be held responsible for maintaining his portion of the status board. (Of course, the instructor should also maintain a careful written record of student progress and task certification.) Prominent display of such a student-maintained status board has markedly beneficial effects on over-all student motivation, behavior, and rate of progress. It is interesting to note, however, that motivation, classroom behavior, and rate of progress seem to reflect directly the instructor's attitude and concern for careful maintenance of the status board.

Several readily apparent benefits accrue from proper use of a class status board:

- increased student motivation
- increased student productivity

POWER MECHANICS
STATUS
BOARD
1-1 SHOP SAFETY
1-2 FIRE SAFETY
1-3 MECHANIC'S HANDTOOLS
1-4 AUTOMOTIVE HARDWARE
1-5 AUTOMOTIVE TERMINOLOGY
2-1 SERVICE-SALES
3-1 JACKS-LIFTS
4-1 WASH-POLISH
5-1 DRIVE BELTS
6-1 MINIATURE BULBS
6-2 HEADLAMPS
7-1 CLEAN-TEST PLUGS
7-2 CHANGING PLUGS
8-1 TIRE INFLATION
8-2 REMOVE-ROTATE WHEELS
8-3 INSPECT TIRES
8-4 TIRE MOUNTING
8-5 REPAIR TIRES-TUBES
9-1 BATTERY HYDROMETER
9-2 INSPECT BATTERIES
9-3 CLEAN BATTERIES
9-4 REMOVE-REPLACE BATTERY
9-5 CHARGING BATTERIES
10-1 RADIATOR PRESSURE CAPS
10-2 TESTING PRESSURE CAPS
10-3 TESTING COOLING SYS.
10-4 COOLING ANTIFREEZE
10-5 INSP. COOLING SYSTEM HOSES
10-6 FLUSHING COOLING SYSTEM
10-7 TESTING COOLING SYS.
10-8 TESTING THERMOSTATS
11-1 REMOVE-REPLACE THERMOSTATS
11-2 LUBRICATION SERVICE
11-3 BODY LUBRICATION
11-4 ENGINE OIL-AFT
11-5 AIR-GAS FILTER
11-6 OIL-FILTER CHANGE
CHASSIS LUBRICATION
ABSENT
Allen, C
Allsopp, R
Beguerie, R
Bowers, D
Burke, T
Coull, D
Deady, T
Dimaggio, D
Hamn, R
Hogan, E
Hunter, L
Johnson, S
OFFICE
LIBRARY

- increased cooperative tutoring among students
- increased instructor-student interaction
- decreased clerical workload for instructor
- decreased instructor anxiety over class management
- decreased student discipline problems

A graphic representation of class progress and accomplishment has a marked effect on both the students and the instructors. Apparently, when properly handled, the status board leads to cooperative behavior rather than to competitive attitudes among the students; and thus the status board comes to stand for class accomplishment rather than individual progress. Most importantly, advancement from learning task to learning task and subsequent graduation into the next higher job level become a visible fact and an attainable goal.

APPENDIX O

SAMPLE EVALUATION AND FOLLOW-UP STUDY FORMS

PROJECT ABLE

INSTRUCTOR REACTION FORM

(PERFORMANCE EVALUATION SET & LEARNER ACTIVITY GUIDE)

INSTRUCTIONS

This checklist is designed to assist in identifying problems in learning units and performance evaluation units. Most items will require only a check mark (√) to give your answer. Please answer all items ACCURATELY. Your comments will be most valuable.

Thanks for your help.

Name ______________________ School ______________ City ______________

Job Family Area and Level ______________________________

Group or Grade ______________________ Date ______________________

Unit Number
- Learning Unit No. ______________
- Performance Evaluation No. ______________

NOTE: YOU MAY CHECK MORE THAN ONE ANSWER.

UNIT OVERALL EVALUATION

☐ The objectives and units are not sequenced correctly (specify).
☐ Requires extensive teacher help.
☐ Needs a greater variety of learning activities.
Reading level within unit too difficult for my students. (Select appropriate one.)
☐ Better ☐ Average ☐ Poor
☐ Please revise as indicated on the attached copy of the unit.
☐ This unit should be deleted from the program. (Why)
☐ There is not enough difference in the units. (How should they be modified?)
☐ The typical student requires too long to complete the unit.
☐ Acceptable as is.
☐ Acceptable with minor revision.

OBJECTIVE

☐ Acceptable.
☐ Needs to be written in simpler language for the student.
☐ Not in correct sequence. (Where should it be?)
☐ Does not tell student what he is supposed to learn.

OVERVIEW

☐ Acceptable.
☐ Needs to be written in simpler language for the student.
☐ Not related to the objective.

INSTRUCTIONAL AND/OR RESOURCE MATERIALS

Where more than one reference is used in the step, indicate which reference a specific comment is directed toward.
☐ Acceptable.
☐ Instructional materials not related to the objective.
☐ Instructional materials require extensive teacher help.
☐ Reading level is too difficult for my students.
☐ Please revise as indicated on the form or on the attached unit.
☐ There is a mistake in page reference, title of book, etc. Correct as indicated on the form or attached unit.
☐ Instructional materials not available in our school. (Which materials?)

EQUIPMENT AND TRAINING AIDS

- ☐ Acceptable.
- ☐ Not related to objectives.
- ☐ Requires too much teacher help.
- ☐ Too difficult for my students.
- ☐ Too dangerous – safety problems (specify).
- ☐ Too difficult to build.
- ☐ Revisions and modifications needed as indicated on form or attached unit.
- ☐ Too difficult or expensive to buy.

TEST QUESTIONS

- ☐ Acceptable.
- ☐ Not related to objectives.
- ☐ Too difficult for my students.
- ☐ Takes too long.
- ☐ Reading and words too difficult.
- ☐ Students dislike them.
- ☐ Revisions needed as indicated on attached unit.

PERFORMANCE ACTIVITY

- ☐ Acceptable.
- ☐ Activities not related to the objective, or they are irrelevant to overall development. (Point out on attached unit.)
- ☐ Objective needs additional activities as indicated on the form or on attached unit in order to prepare students adequately for the achievement of the objective.
- ☐ The activities are not in the correct sequence. Please revise as on the form or attached unit.
- ☐ Activities require extensive teacher help.
- ☐ Too much reading required.
- ☐ Additional activities are needed. (What activities?)
- ☐ Activities are too complicated for students.
- ☐ Activities take too long to complete.
- ☐ There are too many activities.
- ☐ Activities create shop problems. (What problems?)
- ☐ Revisions needed as indicated on attached unit.

STOP______________ Instructor Check
Initials

- ☐ Acceptable.
- ☐ Too frequent.
- ☐ More needed as indicated on attached unit.
- ☐ Please revise as indicated on attached unit.

CRITERION CHECKLIST

- ☐ Acceptable.
- ☐ Needs to be written in simpler language. (Indicate vocabulary or structure causing difficulty.)
- ☐ Does not appear to be related to the objectives.
- ☐ Format is confusing – needed teacher explanation.
- ☐ Insufficient information is given in order to know what is intended. (Specify.)
- ☐ Too much reading – too much detail.
- ☐ Requires too much time for the student.
- ☐ Requires too much time of the instructor.
- ☐ Please revise as indicated on the form or on the attached copy of the checklist.

PROJECT ABLE

STUDENT REACTION FORM

(PERFORMANCE EVALUATION SET AND LEARNER ACTIVITY GUIDE)

INSTRUCTIONS

Most items will require only a check mark (√) to give your answer. Please answer all items ACCURATELY.

Thanks for your help on this important study.

Vocational area ______________________ School ______________ Date__________

Unit Number → Learning Unit No. ______________
→ Performance Evaluation No. ______________

NOTE: YOU MAY CHECK MORE THAN ONE ANSWER.

---

1. Which statements describe the activities in this unit?
   - ☐ Interesting
   - ☐ Easy
   - ☐ Hard
   - ☐ Fun
   - ☐ Too much reading.
   - ☐ Useful
   - ☐ New material (Things you did not know).
   - ☐ Too much theory.

2. Describe the **help** you received on the unit.
   - ☐ I received no help. (Go to number 4.)
   - ☐ I didn't need help.
   - ☐ I received help from another student.
   - ☐ I received help from my instructor.
   - ☐ I received help from others.

   (Go to No. 3.)

3. If you needed help – why?
   - ☐ I was unable to understand what I was to do.
   - ☐ The words were too difficult.
   - ☐ It did not cover what was to be learned.
   - ☐ The objectives did not explain what was to be learned.
   - ☐ The activities were too difficult.
   - ☐ I needed help to locate materials, or tools, or aids, etc.

4. Did you have problems?
   - ☐ following directions?
   - ☐ understanding charts or graphs?
   - ☐ getting supplies or equipment?
   - ☐ using the audio-visuals?
   - ☐ using the training aids?
   - ☐ working on your own?

5. How might we change or improve the unit?

VOCATIONAL EDUCATION IN UNITED STATES

An AMERICAN INSTITUTE FOR RESEARCH Survey

INSTRUCTIONS

Most items on this questionnaire require only a check mark (√) to give your answer. Please answer all items ACCURATELY. The information will be STRICTLY CONFIDENTIAL.

Please return the questionnaire in the postage-paid, pre-addressed envelope provided.

THANKS FOR YOUR HELP ON THIS IMPORTANT STUDY

1. Your Name ______________________ 2. Your High School's Name ______________________

3. Year Graduated from High School: Mo ______ Yr ______ 4. High School Course Studied ______________________

5. Below are ways students are influenced to select a vocational course. Mark those that influenced you to choose the course you took in high school. CIRCLE THE NUMBER OF THE MOST IMPORTANT INFLUENCE

- ☐ 1. Books and magazines
- ☐ 2. Parents
- ☐ 3. Brother or sister
- ☐ 4. Relative
- ☐ 5. Neighbor (adult)
- ☐ 6. Friend your age
- ☐ 7. Job opportunities
- ☐ 8. Part-time job
- ☐ 9. School teacher
- ☐ 10. School counselor
- ☐ 11. School principal
- ☐ 12. Course graduate
- ☐ 13. Other, specify below. ______________________

6. Did your school offer the vocational course you really wanted to take?

- ☐ 1. Yes → If Yes, did you get to take it? → ☐ 6-1.1 Yes, I took the course I wanted.
  - ☐ 6-1.2 No, I could not take the course I wanted → because ______________________
- ☐ 2. No → If No, what course did you want to take that was not offered? ______________________

7. How long after leaving high school did it take you to get your first full-time job? ______ months

8. How did you get your first full-time job after leaving high school? (Mark all that apply.)

- ☐ 1. By answering a want-ad
- ☐ 2. Private employment agency
- ☐ 3. State employment agency
- ☐ 4. Help of school teacher
- ☐ 5. Help of school counselor
- ☐ 6. Help of school principal
- ☐ 7. Help of school placement service
- ☐ 8. Help of friend or relative
- ☐ 9. Through school coop program
- ☐ 10. Other than above

If you never had a full-time job, mark here – ☐ **SKIP TO ITEM 12**

9. Was your first full-time job in the trade or field for which you were trained in high school?

If Yes: Indicate how well your vocational course prepared you for your first full-time job.

- ☐ 1. Exceptionally well-prepared; training covered all essentials required by first job
- ☐ 2. Well-prepared on the whole; but there were some important gaps in training
- ☐ 3. Poorly prepared; much that I needed to know was not covered in vocational course

If No: Mark reason below.

- ☐ 1. No job available in area of training
- ☐ 2. Learned new job by continuing school
- ☐ 3. Learned new job in military service
- ☐ 4. Decided I liked other work better
- ☐ 5. Not accepted as apprentice in trade
- ☐ 6. Other (specify) ______________________

10. How did the: (1) tools and equipment, (2) work methods, and (3) work materials used on your first full-time job compare with those used in your vocational shop courses? If a sub-item is not applicable, mark the box NA to the right. Otherwise, mark your answer.

TOOLS & EQUIPMENT ☐ NA	WORK METHODS ☐ NA	WORK MATERIALS ☐ NA
☐ 1. Identical or almost so	☐ 1. Identical or almost so	☐ 1. Identical or almost so
☐ 2. Little real difference	☐ 2. Little real difference	☐ 2. Little real difference
☐ 3. Very much different	☐ 3. Very much different	☐ 3. Very much different
If you marked 3 above (Very much different), did it take long to learn what was new?		
☐ 1. Only about a few weeks	☐ 1. Only about a few weeks	☐ 1. Only about a few weeks
☐ 2. Less than three months	☐ 2. Less than three months	☐ 2. Less than three months
☐ 3. About three-six months	☐ 3. About three-six months	☐ 3. About three-six months
☐ 4. About six months-a year	☐ 4. About six months-a year	☐ 4. About six months-a year
☐ 5. More than a year	☐ 5. More than a year	☐ 5. More than a year

11.

For each of the skill areas listed below, answer the four questions at the right. → Indicate your answers by marking appropriate boxes.	1 How important is this skill for your present job?				2 How much of this skill was learned in high school?				3 Where did you learn the most about this skill?					4 Do you feel the need for more instruction or training in this area? (Mark either Yes or No)
	1 Of No Real Importance	2 Slightly Important	3 Considerably Important	4 Of Critical Importance	1 Almost Nothing	2 Some, But Not Much	3 Large Amount	4 Almost All	1 High School Coop Program	2 High School Shop or Class	3 Apprentice Program	4 On Regular Job	5 Elsewhere	
1 MANUAL JOB SKILLS. Refers to skill at using or operating tools, equipment, materials, machines, etc., in your work.														1. Yes 2. No
2 JOB PRACTICAL KNOWLEDGE. Refers to practical everyday knowledge of work processes, methods, procedures, etc.														1. Yes 2. No
3 JOB THEORETICAL KNOWLEDGE. Refers to knowledge of basic principles and concepts underlying the practical trade work.														1. Yes 2. No
4 MATHEMATICAL SKILLS. Refers to ability to use arithmetic or higher mathematics to solve work problems.														1. Yes 2. No
5 COMMUNICATION SKILLS. Refers to skill at speaking, writing, drafting, sketching, etc., to communicate ideas.														1. Yes 2. No
6 READING AND INTERPRETIVE SKILLS. Refers to skill at reading printed matter, blueprints, tables, diagrams, etc.														1. Yes 2. No
7 CLERICAL SKILLS. Refers to skill at keeping records, making out reports, and other types of routine paper work.														1. Yes 2. No
8 PERSONAL RELATIONS SKILLS. Refers to skill at dealing with people, such as customers, co-workers, other trades, etc.														1. Yes 2. No
9 SUPERVISORY SKILLS. Refers to skill at supervising others, e.g., instructing, directing, evaluating, planning, organizing, etc.														1. Yes 2. No
10 OTHER SKILLS. Add what you feel applies to your job and is not covered by the above. ______														1. Yes 2. No

12. Please give your frank opinion about the following items concerning your high school education. (Mark one answer for each item.)

ON THE WHOLE:

	1. Poor	2. Satisfactory	3. Good	4. Excellen
1. Quality of instruction from shop instructors				
2. Quality of instruction from academic teachers				
3. Condition of shop facilities and equipment				
4. General physical condition of school				
5. Vocational counseling given to students				
6. Help given students to find jobs				
7. Opportunity for extra-curricular activities				
8. Interest shown by teachers in student problems				
9. Reputation of the school in community				
10. Strictness of school in maintaining discipline				

13. Please mark all kinds of education obtained since leaving high school, and provide the information requested about each. Put an asteris (*) behind those you are presently attending. If you have not had any additional education since high school, mark here → ☐ Estimate your average hours per week over the total period attended.

Mark Here	Type of Education	Major Subject or Course(s)	Dates Attended (Give Month & Year)	Leave Blank	Avg. Hrs. Per Wk. in School	Leave Blank
0	Two-year or junior college		Fr: To:			
1	Four-year college/university		Fr: To:			
2	Post-college graduate school		Fr: To:			
3	Private trade/technical school		Fr: To:			
4	Public trade/technical school		Fr: To:			
5	Business-commercial school		Fr: To:			
6	Adult continuation school		Fr: To:			
7	Military specialist school		Fr: To:			
8	Company course or school		Fr: To:			
9	Correspondence courses		Fr: To:			
10	Other (specify)		Fr: To:			

14. **JOB HISTORY SINCE HIGH SCHOOL.** Start with your **FIRST** job after leaving high school. List **ALL** full-time jobs. List **ONLY** part-time jobs held six months or more, except if your present job is a part-time job. List the jobs in the order that you held them, up to and including your **PRESENT JOB.** (If self-employed, give **NET EARNINGS,** not **GROSS INCOME** of your business.)

	1	2	3	4	5	6	7	8	9	10
EXAMPLE	Starting Date Mo. 7 Yr. 53 Leaving Date Mo. 10 Yr. 54	What type of work did you do? MACHINIST APPRENTICE	Did job require move to new city? 1. ☐ No 2. ☒ Yes How many miles? 20	Full Time 1. ☐ Yes 2. ☒ No If part-time, how many hours per week, on average? 15	Self Employed 1. ☐ Yes 2. ☒ No	Was the work related to vocational course you took? 1. ☒ Same trade studied. 2. ☐ Highly related 3. ☐ Slightly related 4. ☐ Completely unrelated	On the whole, were you satisfied with the work? 1. ☐ Very satisfied 2. ☒ Satisfied 3. ☐ Dissatisfied 4. ☐ Very dissatisfied	Earnings at Starting Give $ per hr., wk., or mo. $ 1.25 per HR Earnings at Leaving Give $ per hr., wk., or mo. $ 1.50 per HR	Reason For Leaving Job NO WORK	Were you unemployed after leaving job? 1. ☒ Yes 2. ☐ No If Yes, how long? 5 MONTHS
1st JOB	Starting Date Mo.___Yr.___ Leaving Date Mo.___Yr.___	What type of work did you do?	Did job require move to new city? 1. ☐ No 2. ☐ Yes How many miles	Full Time 1. ☐ Yes 2. ☐ No If part-time, how many hours per week, on average?	Self Employed 1. ☐ Yes 2. ☐ No	Was the work related to vocational course you took? 1. ☐ Same trade studied 2. ☐ Highly related 3. ☐ Slightly related 4. ☐ Completely unrelated	On the whole, were you satisfied with the work? 1. ☐ Very satisfied 2. ☐ Satisfied 3. ☐ Dissatisfied 4. ☐ Very dissatisfied	Earnings at Starting Give $ per hr., wk., or mo. $ ______ per ______ Earnings at Leaving Give $ per hr., wk., or mo. $ ______ per ______	Reason For Leaving Job	Were you unemployed after leaving job? 1. ☐ Yes 2. ☐ No If Yes, how long?
2nd JOB	Starting Date Mo.___Yr.___ Leaving Date Mo.___Yr.___	What type of work did you do?	Did job require move to new city? 1. ☐ No 2. ☐ Yes How many miles?	Full Time 1. ☐ Yes 2. ☐ No If part-time, how many hours per week, on average?	Self Employed 1. ☐ Yes 2. ☐ No	Was the work related to vocational course you took? 1. ☐ Same trade studied 2. ☐ Highly related 3. ☐ Slightly related 4. ☐ Completely unrelated	On the whole, were you satisfied with the work? 1. ☐ Very satisfied 2. ☐ Satisfied 3. ☐ Dissatisfied 4. ☐ Very Dissatisfied	Earnings at Starting Give $ per hr., wk., or mo. $ ______ per ______ Earnings at Leaving Give $ per hr., wk., or mo. $ ______ per ______	Reason For Leaving Job	Were you unemployed after leaving job? 1. ☐ Yes 2. ☐ No If Yes, how long?
3rd JOB	Starting Date Mo.___Yr.___ Leaving Date Mo.___Yr.___	What type of work did you do?	Did job require move to new city? 1. ☐ No 2. ☐ Yes How many miles?	Full Time 1. ☐ Yes 2. ☐ No If part-time, how many hours per week, on average?	Self Employed 1. ☐ Yes 2. ☐ No	Was the work related to vocational course you took? 1. ☐ Same trade studied 2. ☐ Highly related 3. ☐ Slightly related 4. ☐ Completely unrelated	On the whole, were you satisfied with the work? 1. ☐ Very satisfied 2. ☐ Satisfied 3. ☐ Dissatisfied 4. ☐ Very dissatisfied	Earnings at Starting Give $ per hr., wk., or mo. $ ______ per ______ Earnings at Leaving Give $ per hr., wk., or mo. $ ______ per ______	Reason For Leaving Job	Were you unemployed after leaving job? 1. ☐ Yes 2. ☐ No If Yes, how long?
4th JOB	Starting Date Mo.___Yr.___ Leaving Date Mo.___Yr.___	What type of work did you do?	Did job require move to new city? 1. ☐ No 2. ☐ Yes How many miles?	Full Time 1. ☐ Yes 2. ☐ No If part-time, how many hours per week, on average?	Self Employed 1. ☐ Yes 2. ☐ No	Was the work related to vocational course you took? 1. ☐ Same trade studied 2. ☐ Highly related 3. ☐ Slightly related 4. ☐ Completely unrelated	On the whole, were you satisfied with the work? 1. ☐ Very satisfied 2. ☐ Satisfied 3. ☐ Dissatisfied 4. ☐ Very dissatisfied	Earnings at Starting Give $ per hr., wk., or mo. $ ______ per ______ Earnings at Leaving Give $ per hr., wk., or mo. $ ______ per ______	Reason For Leaving Job	Were you unemployed after leaving job? 1. ☐ Yes 2. ☐ No If Yes, how long?
5th JOB	Starting Date Mo.___Yr.___ Leaving Date Mo.___Yr.___	What type of work did you do?	Did job require move to new city? 1. ☐ No 2. ☐ Yes How many miles?	Full Time 1. ☐ Yes 2. ☐ No If part-time, how many hours per week, on average?	Self Employed 1. ☐ Yes 2. ☐ No	Was the work related to vocational course you took? 1. ☐ Same trade studied 2. ☐ Highly related 3. ☐ Slightly related 4. ☐ Completely unrelated	On the whole, were you satisfied with the work? 1. ☐ Very satisfied 2. ☐ Satisfied 3. ☐ Dissatisfied 4. ☐ Very dissatisfied	Earnings at Starting Give $ per hr., wk., or mo. $ ______ per ______ Earnings at Leaving Give $ per hr., wk., or mo. $ ______ per ______	Reason For Leaving Job	Were you unemployed after leaving job? 1. ☐ Yes 2. ☐ No If Yes, how long?
6th JOB	Starting Date Mo.___Yr.___ Leaving Date Mo.___Yr.___	What type of work did you do?	Did job require move to new city? 1. ☐ No 2. ☐ Yes How many miles?	Full Time 1. ☐ Yes 2. ☐ No If part-time, how many hours per week, on average	Self Employed 1. ☐ Yes 2. ☐ No	Was the work related to vocational course you took? 1. ☐ Same trade studied 2. ☐ Highly related 3. ☐ Slightly unrelated 4. ☐ Completely unrelated	On the whole, were you satisfied with the work? 1. ☐ Very satisfied 2. ☐ Satisfied 3. ☐ Dissatisfied 4. ☐ Very dissatisfied	Earnings at Starting Give $ per hr., wk., or mo. $ ______ per ______ Earnings at Leaving Give $ per hr., wk., or mo. $ ______ per ______	Reason For Leaving Job	Were you unemployed after leaving job? 1. ☐ Yes 2. ☐ No If Yes, how long?

ATTENTION: If you held more than six full and part-time jobs, please continue on the page enclosed. Be sure to include your present full-time and/or part-time job. Thank you.

15. **YOUR PRESENT JOB.** (Please give this additional information.)

1. Present Earnings? Give $ per hour, week, or month. $ ______________ per __________
2. Your Employer: ______________________________

   Street Address: ______________________________

   City-State: ______________________________

16. Did you have any military service? ☐ 1. No ☐ 2. Yes ⟶ How many months? ________ Nature of work? ________

17. Were you unemployed for reason of health or hospitalization? ☐ 1. No ☐ 2. Yes ⟶ How many months? ________

Part of our study concerns the interests, activities, and associations of high school graduates. We hope you will not regard this information too personal to give us. All is confidential. Please weigh your answers carefully.

18. How frequently do you talk about the following topics when you get together socially with others?

	ALMOST NEVER 1	INFREQUENTLY 2	FREQUENTLY 3	ALMOST ALWAYS 4
1. Your work	☐	☐	☐	☐
2. Religion	☐	☐	☐	☐
3. Politics	☐	☐	☐	☐
4. Business conditions	☐	☐	☐	☐
5. World affairs	☐	☐	☐	☐
6. National affairs	☐	☐	☐	☐
7. State affairs	☐	☐	☐	☐
8. Community problems	☐	☐	☐	☐
9. Your hobbies	☐	☐	☐	☐
10. Sports and athletics	☐	☐	☐	☐
11. Music, art, literature, etc.	☐	☐	☐	☐
12. Government matters	☐	☐	☐	☐
13. Labor union matters	☐	☐	☐	☐
14. Your family	☐	☐	☐	☐
15. Other (specify) ________	☐	☐	☐	☐

19. How frequently do you engage in the following types of leisure-time activities?

	ALMOST NEVER 1	INFREQUENTLY 2	FREQUENTLY 3	ALMOST DAILY 4
1. Reading newspapers	☐	☐	☐	☐
2. Engaging in craft hobbies (model building, jewelry making, etc.)	☐	☐	☐	☐
3. Reading professional or trade books and periodicals	☐	☐	☐	☐
4. Attending athletic events as a spectator	☐	☐	☐	☐
5. Attending plays, concerts, ballets, etc.	☐	☐	☐	☐
6. Watching television programs	☐	☐	☐	☐
7. Gardening (raising flowers, fruit trees, vegetables, etc.)	☐	☐	☐	☐
8. Reading general magazines (LIFE, LOOK, READERS' DIGEST, etc.)	☐	☐	☐	☐
9. Working at home shop activities (woodworking, metalworking, etc.)	☐	☐	☐	☐
10. Attending educational courses for self-improvement	☐	☐	☐	☐
11. Engaging in team sports (softball, football, etc.)	☐	☐	☐	☐
12. Engaging in performing arts (acting, singing, instruments, etc.)	☐	☐	☐	☐
13. Visiting or entertaining friends	☐	☐	☐	☐
14. Reading non-fiction books (biography, history, travel, etc.)	☐	☐	☐	☐
15. Collecting stamps, coins, rocks, or other items	☐	☐	☐	☐
16. Attending educational lectures and discussion groups	☐	☐	☐	☐
17. Engaging in individual sports (swimming, hunting, fishing, etc.)	☐	☐	☐	☐
18. Listening to music at home for pleasure	☐	☐	☐	☐
19. Going to the movies	☐	☐	☐	☐
20. Other (please write in) ________	☐	☐	☐	☐

20. Below is a list of different type organizations and associations. Mark the space which best describes your membership status in each type of organization, association, or club.

	NOT A MEMBER 1	INACTIVE MEMBER 2	ACTIVE MEMBER 3	PRESENTLY AN OFFICER 4
1. A church or a religious organization	☐	☐	☐	☐
2. Political organization	☐	☐	☐	☐
3. Service organization (Rotary, Lions, Kiwanis, etc.)	☐	☐	☐	☐
4. Sports club or athletic organization	☐	☐	☐	☐
5. Labor union	☐	☐	☐	☐
6. Fraternal organization (Elks, Masons, K. of C., etc.)	☐	☐	☐	☐
7. Veterans' organization	☐	☐	☐	☐
8. Business or trade association	☐	☐	☐	☐
9. Music or other cultural association	☐	☐	☐	☐
10. Local civic association	☐	☐	☐	☐
11. Youth organization (Scouts, Y.M.C.A., etc.)	☐	☐	☐	☐
12. Professional association	☐	☐	☐	☐
13. Other (specify) ________	☐	☐	☐	☐

21. Marital Status

☐ 1. Single
☐ 2. Married
☐ 3. Other

22. Race

☐ 1. White
☐ 2. Negro
☐ 3. Other

23. Religion

☐ 1. Protestant
☐ 2. Catholic
☐ 3. Jewish
☐ 4. Other
☐ 5. None

24. Do you have any disability or health condition that limits your employability?

☐ 1. Yes ☐ 2. No

THANK YOU FOR YOUR TIME AND EFFORT

PROJECT ABLE
VOCATIONAL GRADUATE FOLLOW–UP

INSTRUCTIONS

Most items on this questionnaire require only a check mark (√) to give your answer. Please answer all items ACCURATELY. The information will be STRICTLY CONFIDENTIAL.

Please return the questionnaire in the postage-paid, pre-addressed envelope provided.

THANKS FOR YOUR HELP ON THIS IMPORTANT STUDY

Company or firm ______________________ Supervisors of employee ______________________

Employee's Name ______________________ Your Name ______________________ Date ________

Please evaluate the person in question in terms of the characteristics indicated below by checking the appropriate spaces.

For each of the skill areas listed below, answer the four questions at the right. → Indicate your answers by marking appropriate boxes.	1 How important is this skill for his present job				2 How would you evaluate him on this skill				3 How does he compare with others of about his age who had other training?				4 Does he need more instruction or training in this area? (Mark either Yes or No)
	1 Of no real importance	2 Slightly important	3 Considerably important	4 Of critical importance	1 Needs much improvement	2 Generally satisfactory	3 Generally above average	4 Outstanding	1 Have no one to fairly compare him with	2 Does not do as well as others doing same work	3 Does about as well as others doing same work	4 Does better than others doing same work	
1. **MANUAL JOB SKILLS.** Refer to skill at using or operating tools, equipment, materials, machines, etc., in your work.	☐	☐	☐	☐	☐	☐	☐	☐	☐	☐	☐	☐	☐ 1. Yes ☐ 2. No
2. **JOB PRACTICAL KNOWLEDGE.** Refers to practical everyday knowledge of work processes, methods, procedures, etc.	☐	☐	☐	☐	☐	☐	☐	☐	☐	☐	☐	☐	☐ 1. Yes ☐ 2. No
3. **JOB THEORETICAL KNOWLEDGE.** Refers to knowledge of basic principles and concepts underlying the practical trade work.	☐	☐	☐	☐	☐	☐	☐	☐	☐	☐	☐	☐	☐ 1. Yes ☐ 2. No
4. **MATHEMATICAL SKILLS.** Refers to ability to use arithmetic or higher mathematics to solve work problems.	☐	☐	☐	☐	☐	☐	☐	☐	☐	☐	☐	☐	☐ 1. Yes ☐ 2. No
5. **COMMUNICATION SKILLS.** Refers to skill at speaking, writing drafting, sketching, etc., to communicate ideas.	☐	☐	☐	☐	☐	☐	☐	☐	☐	☐	☐	☐	☐ 1. Yes ☐ 2. No
6. **READING AND INTERPRETIVE SKILLS.** Refers to skill at reading printed matter, blueprints, tables, diagrams, etc.	☐	☐	☐	☐	☐	☐	☐	☐	☐	☐	☐	☐	☐ 1. Yes ☐ 2. No
7. **CLERICAL SKILLS.** Refers to skill at keeping records, making out reports, and other types of routine paper work.	☐	☐	☐	☐	☐	☐	☐	☐	☐	☐	☐	☐	☐ 1. Yes ☐ 2. No
8. **PERSONAL RELATIONS SKILLS.** Refers to skill at dealing with people, such as customers, co-workers, other tradesmen, etc.	☐	☐	☐	☐	☐	☐	☐	☐	☐	☐	☐	☐	☐ 1. Yes ☐ 2. No
9. **SUPERVISORY SKILLS.** Refers to skill at supervising others, e.g., instructing, directing, evaluating, planning, organizing etc.	☐	☐	☐	☐	☐	☐	☐	☐	☐	☐	☐	☐	☐ 1. Yes ☐ 2. No
10. **ATTITUDE TOWARD WORK.** Refers to such behavior as absenteeism, rule violation, concern for quality work, cooperation, etc.	☐	☐	☐	☐	☐	☐	☐	☐	☐	☐	☐	☐	☐ 1. Yes ☐ 2. No

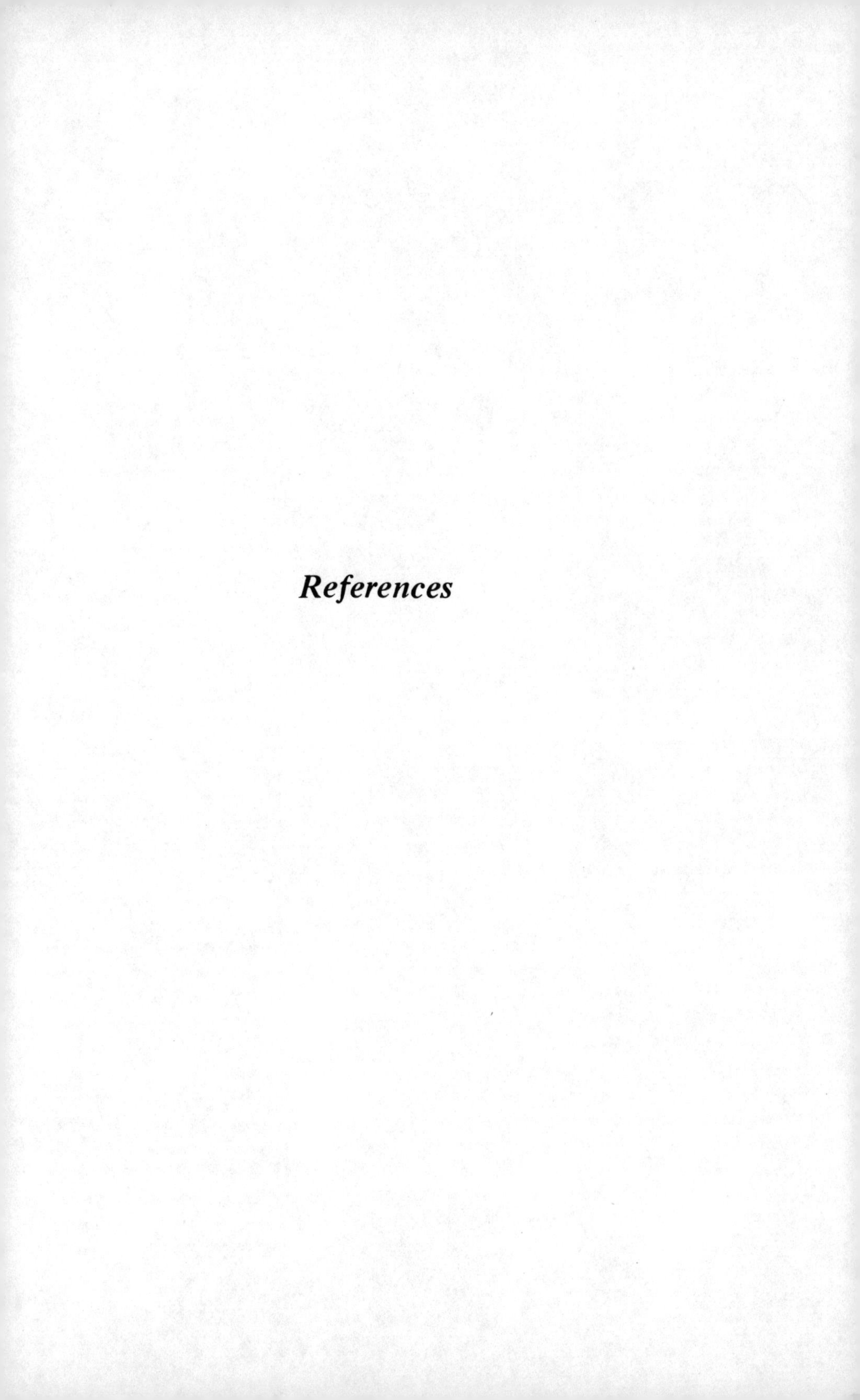

References

REFERENCES

Altman, J.W. 1966. *Research on General Vocational Capabilities (Skills and Knowledges)* Pittsburgh: American Institutes for Research.

________ & Morrison, E.J. 1966. *School and Community Factors in Employment Success of Trade and Industry Course Graduates.* Pittsburgh: American Institutes for Research.

American Institutes for Research. 1964. An Experimental Program. *Development and Evaluation of an Experimental Curriculum for the New Quincy (Mass.) Vocational-Technical School.* Pittsburgh: Institute for Performance Technology.

________. *Development and Evaluation of an Experimental Curriculum for the New Quincy (Mass.) Vocational-Technical School.* Pittsburgh: Institute for Performance Technology:

First Quarterly Technical Report. 1965.

The Problem of Defining Objectives. 1965. Second quarterly technical report.

Curriculum Implications of the Study of Objectives. 1965. Third quarterly technical report.

A Vocational Guidance Program for Junior High School Students. 1966. Fourth quarterly technical report.

The Roles, Characteristics and Development Procedures for

Measurement of Individual Achievement. 1966. Fifth quarterly technical report.

The Development of Learning Units. 1966. Sixth quarterly technical report.

The Sequencing of Learning Units. 1966. Seventh quarterly technical report.

Problems Relating to the Development and Implementation of a Vocational Curriculum. 1967. Eighth quarterly technical report.

Development and Tryout of a Junior High School Student Vocational Guidance Plan. 1967. Ninth quarterly technical report.

A Mathematics Curriculum for Vocational Education. 1968. Tenth quarterly technical report.

The Electronics Curriculum. 1968. Eleventh quarterly technical report.

The Power Mechanics Curriculum. 1969. Twelfth quarterly technical report.

Management and Evaluation Plan for Instructional Systems Development for Vocational-Technical Education. 1970. Fifteenth quarterly technical report.

The Woodworking Curriculum. 1970. Sixteenth quarterly technical report.

Atkin, J.M. 1963. Some Evaluation Problems in a Course Content Improvement Project. *Journal of Research in Science Teaching,* Vol. 1, No. 2.

________. 1968. Behavioral Objectives in Curriculum Design.

Science Teaching, Vol. 35, No. 5.

Ausubel, D.P. 1966. *Psychological Aspects of Curriculum Evaluation.* Paper read at the National Seminar for Research and Curriculum Evaluation in Vocational and Technical Education, University of Illinois.

Baker, Eva. 1966. *Establishing Performance Standards.* Los Angeles: Vimcet Associates.

Bilodeau, E.A. 1966. *Acquisition of Skill.* New York: Academic Press.

Bloom, B.S. *et al.* 1956. *Taxonomy of Educational Objectives: The Classification of Educational Goals, Handbook I: Cognitive Domain.* New York: David McKay Co.

_______ *et al.* 1971. *Handbook of Formative and Summative Evaluation.* New York: McGraw-Hill.

Briggs, L.J. 1967. *Sequencing of Instruction in Relation to Hierarchies of Competence.* Palo Alto, California: American Institutes for Research.

_______, Campeau, P.L., Gagné, R.M. & May, M.A. 1967. *Instructional Media: A Procedure for the Design of Multi-Media Instruction, a Critical Review of Research, and Suggestions for Future Research.* Pittsburgh: American Institutes for Research.

Broudy, Harry S. 1966. *Problems and Prospects in Vocational Education.* Paper read at the National Seminar for Research and Curriculum Evaluation in Vocational and Technical Education, University of Illinois.

Bruner, J.S. 1966. *Toward a Theory of Instruction.* Cambridge, Mass.: Belknap Press.

Bushnell, David S. 1969. An Educational System for the '70's. *Phi Delta Kappan.*

Butler, F.C. 1963. *A Programmed Television Course in Basic Spanish.* Colorado: Faculty Research Report, U.S. Air Force Academy.

_______. 1964. *A Programmed Television Course in Basic Typing.* Colorado: Faculty Research Report, U.S. Air Force Academy.

_______. 1964. Programmed Instruction and Instructional Systems. *Journal of Training Directors.* American Society of Training Directors.

_______. 1967. *Handbook for Job Corps Instructional Systems Development.* Washington, D.C.: U.S. Government Printing Office.

_______. 1968. *Objectives for Occupational Education.* The National Assessment Program. Pittsburgh: American Institutes for Research.

_______. 1970. *A Rationale and Operating Plan for a Contingency Managed Independent Learning School.* Prepared for Independent Learning Systems, Inc., San Rafael, California.

_______. 1971. *Self-Study Systems–An Instructional Management System for Individualized Mathematics.* New York: Learning Research Associates, Inc.

_______. 1971. *Electronics Technology: An Individualized Job Cluster Career Training Curriculum.* Farmingdale, New Jersey: Education Division (Lab-Volt), Buck Engineering Co.

_______ & Stieger, A.R. 1964. Programmed Instruction and Instructional Systems. In Ofiesh, G.D. & Meierhenry, W.C. (Eds.), *Trends in Programmed Instruction*, published by DAVI-NSPI.

_______ & Hudak, V.M. 1967. *Development and Tryout of a Junior High School Student Vocational Guidance Plan.* Pittsburgh: American Institutes for Research.

_______ & Loch, C. 1968. *An Individualized Mathematics Program for Vocational Education.* Pittsburgh: American Institutes for Research.

_______ & Crozier, P.W. 1968. *An Individualized Electronics Curriculum.* Pittsburgh: American Institutes for Research.

_______, Hudak, V.M. & Kowal, B. 1968. *Problems Relating to the Development and Implementation of an Experimental Vocational Curriculum.* Pittsburgh: American Institutes for Research.

Cook, Desmond L. 1964. *A New Approach to the Planning and Management of Educational Research.* The PERT project, School of Education, The Ohio State University (mimeographed).

Cronbach, L.J. 1963. Course Improvement Through Evaluation. *Teacher's College Record,* Vol. 64.

________. 1966. The Logic of Experiments on Discovery. In Schulman, L.M. & Keislar, E. (Eds.), *Learning by Discovery.* Chicago: Rand-McNally.

Eninger, M.U. 1965. *The Process and Product of T&I High School Level Vocational Education in the United States.* Pittsburgh: American Institutes for Research.

Flanagan, J.C., Davis, F.B., Dailey, J.T., Shaycoft, M.F., Orr, D.B., Goldbert, I. & Neyman, C.A., Jr. 1964. *The American High-School Student.* Washington, D.C.: University of Pittsburgh Project TALENT Office.

________ & Jung, S.M. 1970. *Evaluating a Comprehensive Educational System.* Paper read at the American Institutes for Research Seminar on Evaluative Research, Washington, D.C.

Gage, N.L. 1962. *Handbook of Research on Teaching.* Chicago: Rand-McNally.

Gagné, R.M. (Ed.). 1962. *Psychological Principles in System Development.* New York: Holt, Rinehart, & Winston.

________. 1965. *The Conditions of Learning.* New York: Holt, Rinehart, & Winston.

________ (Ed.). 1966. *Learning and Individual Differences.* Columbus, Ohio: Charles E. Merrill Books.

Gardner, J.W. 1960. National Goals in Education. In *Goals for Americans. The Report of the President's Commission on National Goals.* Englewood Cliffs, N.J.: Prentice-Hall.

Gilbert, T.F. 1962. Mathetics: The Technology of Education. *Journal of Mathetics,* Vol. 1, 7-73.

Glaser, R.L. 1963. *Instructional Technology and the Measurement of Learning Outcomes.* American Psychologist, Vol. 18, 519-522.

________, Damrin, Dora E. & Gardner, F.M. 1952. *The TAB Item: A Technique for the Measurement of Proficiency in Diagnostic Problem-Solving Tasks.* Urbana: Bureau of Research and Service, College of Education, University of Illinois.

________ (Ed.). 1965. *Teaching Machines and Programmed Learn-*

ing, II: Data and Directions. Washington, D.C.: National Education Association, pp. 21-65.

Gropper, George L. & Short, J.G. 1969. *Handbook for Training Development.* Pittsburgh: American Institutes for Research.

Hilgard, E.R. (Ed.). 1964. *Theories of Learning and Instruction.* Sixty-Third Yearbook, National Society for the Study of Education. Chicago: National Society for the Study of Education.

Hudak, V.M. & Morrison, E.J. 1966. *A Vocational Guidance Program for Junior High School Students.* Pittsburgh: American Institutes for Research.

——— & Butler, F.C. 1967. *Development and Tryout of Junior High School Student Vocational Guidance Plan.* Pittsburgh: American Institutes for Research.

Kapfer, Miriam B. (Ed.). 1971. *Behavioral Objectives in Curriculum Development.* Englewood Cliffs, New Jersey: Educational Technology Publications.

Klausmeir, H.J. (Ed.). 1966. *Conceptual Learning.* New York: Academic Press, pp. 81-95.

Krathwohl, D.R., Bloom, B.S. & Masia, B.B. 1964. *Taxonomy of Educational Objectives: The Classification of Educational Goals, Handbook II: Affective Domain.* New York: David McKay Co.

Krumboltz (Ed.). 1965. *Learning and the Educational Process.* Chicago: Rand-McNally, pp. 1-24.

Leporini, F., Neifing, G. & Ullery, J.W. 1970. *A Woodworking Curriculum.* Pittsburgh: American Institutes for Research.

Leton, Donald. 1966. *Criterion Problems and Curriculum Evaluation.* Paper read at the National Seminar for Research and Curriculum Evaluation in Vocational and Technical Education, University of Illinois.

Lexnor, W.B. & Morrison, E.J. 1966. *The Roles, Characteristics, and Development Procedures for Measurement of Individual Achievement.* Pittsburgh: American Institutes for Research.

———. 1966. *The Development of Learning Units.* Pittsburgh: American Institutes for Research.

———. 1966. *The Sequencing of Learning Units.* Pittsburgh:

American Institutes for Research.

Lindvall, C.M. (Ed.). 1964. *Defining Educational Objectives.* Pittsburgh: University of Pittsburgh Press.

Lipham, James M. 1966. *Administrators and the Educational Program.* Paper read at the National Seminar for Research and Curriculum Evaluation in Vocational and Technical Education, University of Illinois.

Lumsdaine, A.A. (Ed.). 1961. *Teaching Machines and Programmed Instruction.* Washington, D.C.: National Education Association.

Mager, Robert F. 1962. *Preparing Objectives for Programmed Instruction.* San Francisco: Fearon.

_______ & McCann, J. 1961. *Learner-Controlled Instruction.* Palo Alto, California: Varian Associates.

Morrison, E.J. 1965. *First Quarterly Report.* Pittsburgh: American Institutes for Research.

_______. 1965. *The Problem of Defining Objectives.* Pittsburgh: American Institutes for Research.

_______. & Gagné, R.M. 1965. *Curriculum Implications of the Study of Objectives.* Pittsburgh: American Institutes for Research.

Popham, W.J. 1967. *Educational Criterion Measures.* Inglewood, California: Southwest Regional Laboratory for Educational Research and Development.

_______. 1969. Objectives of Instruction. In Popham, W.J., Eisner, E.W., Sullivan, H.J. & Tyler, L.L. (Eds.) *Instructional Objectives.* AREA monograph series on curriculum evaluation, No. 3, Chicago: Rand-McNally.

_______. 1969. *Validation Results: Performance Tests of Teaching Proficiency in Vocational Education.* Paper presented at the American Educational Research Association Meeting, Los Angeles.

_______ & Baker, E.L. 1966. *Development of Performance Test of Teaching Proficiency.* Paper presented at the annual meeting of the American Educational Research Association, New York.

_______ & Husek, T.R. 1969. Implications of Criterion-Referenced Measurement. *Journal of Educational Measurement,* Vol. 6, No. 1, pp. 1-9.

——— *et al.* 1971. *Criterion-Referenced Measurement (An Introduction).* Englewood Cliffs, New Jersey: Educational Technology Publications.

Quirk, C. & Sheehan, C. (Eds.). 1967. *Research in Vocational and Technical Education.* Madison: University of Wisconsin: Center for Studies in Vocational and Technical Education.

Rahmlow, Harold F. 1969. *Use of Student Performance Data for Improvement of Individualized Instructional Materials.* Paper read at the American Psychological Association meeting in Washington, D.C.

Rosenfeld, Michael. 1967. *An Evaluation Plan for the Greene Joint Vocational School.* Pittsburgh: American Institutes for Research.

———, Kowal, B. & Selier, E.L. 1965. *Assessing the Progress of Education: Vocational Education–Phase I.* Pittsburgh: American Institutes for Research.

Stake, R.E. 1966. *The Countenance of Education Evaluation.* Paper read at the National Seminar for Research and Curriculum Evaluation in Vocational and Technical Education, University of Illinois.

Tyler, R.W. 1964. *Constructing Achievement Tests.* Columbus: Ohio State University.

———, Gagné, R.M. & Scriven, M. 1967. *Perspectives of Curriculum Evaluation.* AERA monograph series on curriculum evaluation, No. 1. Chicago: Rand-McNally.

Ullery, J.W. 1970. *Management and Evaluation Plan for Instructional Systems Development for Vocational-Technical Education.* Pittsburgh: American Institutes for Research.

——— & Forsyth, R.W. 1969. *The Power Mechanics Curriculum.* Pittsburgh: American Institutes for Research.

U.S. Office of Education. 1963. *Education for a Changing World of Work.* Washington, D.C.: U.S. Government Printing Office.

U.S. Department of Labor. 1965. *Dictionary of Occupational Titles, Volume II: Occupational Classification and Industry Index.* Washington, D.C.: U.S. Government Printing Office.

———. 1965. *Manpower Report of the President and a Report on Manpower Requirements, Resources, Utilization, and Training.* Washington, D.C.: U.S. Government Printing Office.

Venn, G. 1964. *Man, Education, and Work.* Washington, D.C.: American Council on Education.

Index

INDEX